NEIGHBORHOOD TECHNOLOGIES

NEIGHBORHOOD TECHNOLOGIES

MEDIA AND MATHEMATICS OF DYNAMIC NETWORKS

EDITED BY

TOBIAS HARKS AND SEBASTIAN VEHLKEN

DIAPHANES

FIRST PRINTING

ISBN 978-3-03734-523-8

LAYOUT AND PREPRESS: 2EDIT, ZURICH
PRINTED IN GERMANY

WWW.DIAPHANES.NET

TABLE OF CONTENTS

IV. NEIGHBORHOOD ACTIVITIES

ACKNOWLEDGEMENTS

This publication is condensed from the papers read at the Blankensee Colloquium 2012: *Neighborhood Technologies. Media and Mathematics of Dynamic Networks* which took place at the *Denkerei* at Oranienplatz in Berlin/Kreuzberg. It was funded by the Wissenschafts-kolleg/Institute for Advanced Study, *Berlin*. The editors thank the Institute for the support and amenities that it also contributed to this book project. Furthermore, we give thanks to the MECS – Institute for Advanced Study on Media Cultures of Computer Simulation, Leuphana University Lüneburg, the Leuphana Research Service, and the Marie-Curie-Fellowship Program for additional financial funding.

We are indebted to the following people whose personal engagement made the conference and the publication a successful, productive, and lasting experience: Our assistants Mayka Kmoth and Johannes Fuss at the venue, and MECS research student Clara Lotte Warnsholdt for copy-editing the volume, Martin Garstecki and Larissa Buchholz at the Wissenschaftskolleg Berlin, Thomas Eißler and Marina Sawall at the Denkerei, Angelika Stadler for the design and layout of the conference's posters and information sheets, the team of Hotel Johann (Kreuzberg), Claus Pias at MECS, Thomas Puls, Dörte Krahn, Sascha Ludenia, Susanne Falk and Matthias Becker at Leuphana, and Michael Heitz at diaphanes. We also thank Dirk Helbing for granting us the copyright for the publication of his article in this volume. Finally, we want to express our gratitude to all conference speakers and authors who were willing to take part in our transdisciplinary endeavor. May the Force be with you.

SEBASTIAN VEHLKEN, TOBIAS HARKS

NEIGHBORHOOD TECHNOLOGIES
AN INTRODUCTION

> Caught in the middle of a sea of hazy light and a sea of hazy noise, Kate Schechter stood and doubted. All the way out of London to Heathrow she had suffered from doubt. She was not a superstitious person, or even a religious person. She was simply someone who was not at all sure she should be flying to Norway. But she was finding it increasingly easy to believe that God, if there was a God, and if it was remotely possible that any godlike being who could order the disposition of particles at the creation of the Universe would also be interested in directing traffic on the M4, did not want her to fly to Norway either. All the trouble with the tickets, finding a next-door neighbour to look after the cat, then finding the cat so it could be looked after by the next-door neighbour, the sudden leak in the roof, the missing wallet, the weather, the unexpected death of the next-door neighbour, the pregnancy of the cat – it all had the semblance of an orchestrated campaign of obstruction which had begun to assume godlike proportions.
>
> Douglas Adams, *The Long, Dark Tea-Time of the Soul*, p. 2.

1.

When the protagonists of Douglas Adams' novels set off for new adventures, they more often than not are immediately confronted with utterly complex situations. And sometimes – regardless of them being religious or not – they indeed have to assume godlike campaigns behind these events. Even if Kate Schechter's abovementioned attempt to deploy her next-door neighbor as a watchdog for her cat obviously resulted in aggravating complications, contemporary *Neighborhood Technologies* are applied to account for, to control, and to operationalize complex real-life phenomena or social systems in a variety of fields – reaching as far as to the introduction of novel laws of physics in computer simulations (see the articles of Dirk Helbing and Sándor Fekete). These *Neighborhood Technologies* have less to do with social or cultural techniques of building local communities – like greeting the new couple that moved in next door with bread and salt or asking the neighbors to look after one's precious cat or *ficus benjamina* while abroad, let alone the unfathomable field of dealing with neighborhood conflicts. Neither do they primarily allude to architectural or urbanistic concepts of arranging buildings into functioning cityscapes or to describing neighborhoods as a societal unit of community-building. Conversely, the idea for this publication is based on the observation of its editors – one with a background of applied mathematics and computer science, the other in media history and cultural theory – that neighborhoods not only play a substantial role in our respective fields of study. We are also convinced that the transdisciplinary coalescence of a mathematical-computational and a media-historical and -theoretical approach on *Neighborhood Technologies* yields substantial benefits for understanding current attempts to operationalize, describe, experiment with, and theorize complex real-life phenomena.

Whether our contemporary society and culture may be characterized by a mere preoccupation with concepts of space or a predominant fixation on the dynamics of time – with the speculative power of world-wide financial markets and their respective financial tools (e.g., micro-trading) and the increasing penetration of scientific research by computational tools like computer simulations and their implications for a futurologic governmental style (e.g., in fields like climate research or *pre-emption*) as only two protruding pillars: The *time-critical* dynamics of our networked societies are eminently dependent on spacial orderings – be it, just to stick to the above examples, the actual proximity of micro-trading servers to the stock exchange or be it the layout of supercomputing hardware like the placement of chips, cooling systems, and wiring infrastructure. And nevertheless, they can only be studied and understood in a four-dimensional perspective which includes their unpredictable system behavior over time. Oftentimes, these time-critical dynamics can best be analyzed, described, and designed by a topology which is based on local interactions in neighborhoods of similar elements.

On first sight, this observation may sound neither very new nor notably creative: Haven't self-organizing, *bottom-up systems* of simple elements and their nonlinear interactions been a matter of research in various disciplines at least since the boom of complexity science in the 1980s? And doesn't the mathematical interest in *small-world networks* date back at least to the mid-1990s? Granted, but we believe that an examination of *Neighborhood Technologies* still yields interesting results if it does not stop at *analyzing* and *describing* dynamic networks as small worlds or – rather mathematically abstract – self-organizing systems, but takes into focus the application of such a neighborly understanding of network dynamics to real world dynamics. Thus, we are interested in delving into the implementation of (mathematical) neighborhood concepts in operative media technologies and to summon contributions which exemplify the ineluctable transdisciplinary approach that is inherent in such concepts and applications alike. The publication ventures to conduct this examination not only structurally – that is, with contributions from a variety of fields – but also historically, that is, from a selection of applications which not only demonstrate the immediacy, but also the genealogy of *Neighborhood Technologies*.

One early and guiding example of such an entwined mathematical and media-technological understanding of neighborhoods dates back to the year 1971 when the economist, expert on nuclear strategy and later Nobel laureate Thomas C. Schelling published his research on housing segregation in major US-American cities. With his influential paper on *Dynamic Models of Segregation*, Schelling accomplished more than just contributing to a novel type of social mathematics. His interest in the mechanisms of social segregation and its respective models amalgamated the analysis of actual neighborhood dynamics with a *neighborly* research method: Starting from some basic local – a.k.a. neighborly – micro-relations of a defined number of agents whose actions were executed in accordance with a restricted rule set, Schelling dynamically generated macroscopic segregation patterns. The most striking observation of his dynamic models consisted of the fact that these macroscopic patterns generated novel insights which were not deducible from the microscopic properties. For instance, segregation phenomena in urban environments did not necessarily correlate with political attitudes, but were often simply the effect of individuals choosing to live next to similar neighbors. Henceforward, neighborhoods constituted a new research paradigm in which the complex macro-behaviors of a system and the non-linear dynamics of social collectives were *generatively* and *procedurally* put forth by rigidly defined microscopic neighborhood relations. These were, in Schelling's case, first executed on paper, then on checkerboards, and later by computer simulations on cellular automata. The compelling effect was threefold: Neighborhoods at the same time became an *object*, a *conceptual principle*, and a *media technology* for understanding the dynamics of complex real-world phenomena – here, of housing segregation. This provided the basis for the contemporary operationality of *Neighborhood Technologies*. Today, a large number of media-technological applications take advantage of the intermediate level of locally defined, self-coordinating neighborhoods as a *mesoscopic range* to better understand and control the relationship between the interactions of individual agents and the overall global dynamics of complex phenomena. And their fields of application stretch – among others – from Game Theory to Economics, to Sociology and Biology, to Epidemics and Logistics, and to Robotics or Neurology.

2.

The ambiguity of neighborhoods as a contemporary scientific object and operational application plays an important role in mathematical optimization and algorithmic game theory. In optimization one employs the notion of neighborhoods to define local search methods to efficiently compute good solutions for computationally hard optimization problems. In game theory, neighborhood relations are relevant for both the methodology employed (unilateral deviations of players define a neighborhood and thus natural myopic improvement dynamics can be interpreted as executing a local search algorithm) as well as the actual model under investigation (game theoretical analysis of social networks). These disciplines for their part search for actual objects and systems where neighborhood relations play an important role in order to subject them to mathematical analysis. Their major focus lies on predicting, evaluating and qualitatively assessing the state of an uncontrolled system that is determined by distributed actions of (rationally behaved) individuals based on their available information.

The following research questions reflect the major streams of research in this field: Do the actions of individuals guide the system eventually to a stable state (Equilibrium Existence)? How long does it take to reach a stable state by myopic actions of individuals (Convergence of Learning Processes)? What is the complexity of computing or predicting future states of the system? What is the quality of the system at any point in time with respect to a predefined social objective? To which extent can a designer implement rules of interactions so as to drive the system into a desirable state? How vulnerable or manipulable is a system if a group of individuals (for instance, a flash mob, a Facebook campaign by political parties, spam-mail-clients or viruses) coordinate their actions? Or, how well can simulations help predict the system state over time?

Likewise, for some years a growing interest in neighborhood-induced effects can be discovered in certain strands of media theory and history. Be it – to mention only two examples – the ongoing discourse on *swarm intelligence* and the role of distributed (online) communication networks for socio-political action, be it a media-historical approach to *local-based media* (e.g., GPS navigation) and their influence on a transformation of concepts of space and time. Neighborhoods come to be part of not only a topographical and topological, but also a conceptual transformation. As techno-social groupings based on autonomous local interactions, such dynamized neighborhoods become an influential driving force of (global) mass movements. And at the same time, neighborhoods themselves transform into eminently media-technological arrangements, into a particular time-based form of organizing dynamic networks, remodeled according to mathematical conceptions of neighborhoods. In this sense, neighborhoods convert from mere geo-spatial and architectural to media-technological phenomena which manifest as the central hub for theorizing and conceptualizing dynamic networks and their socio-political effects.

As a consequence, typical research questions in the cultural and media studies consider the democratic potential and altered hierarchy levels inherent in techno-social networks, and thus the future modes of socio-political participation. They ask about the relation of "pattern and purpose" in collective human behavior, and how the topologies of locally-organized neighborhood technologies differ from precedent forms of network structures. They inquire into the constitution of neighborhoods between human agents and a variety of non-human tools, applications, and things in shared media-technological environments, and they engage with the epistemological status of emergent phenomena and complexity levels which derive from the micro-behaviors in neighborhoods. And not least, they inquire how, in contemporary agent-based computer simulations, neighborly principles develop into a media technology that provides novel potentials to address a wide variety of real-life phenomena from a bottom-up perspective.

The understanding of neighborhoods in a mathematical and a media-technological sense thus are intrinsically entwined. The theoretical considerations of neighborhood effects and various mathematical and game-theoretical models are the basis for media-technological applications which re-structure the explanatory modes of cultural and societal developments and the production modes of actual architectural or *digital* neighborhoods. These, on the other hand, produce novel demands and challenges for re-structuring and modifying the mathematical tools.

Furthermore, the editors have long believed – even before we started working together on *Neighborhood Technologies* – that a fruitful exchange of ideas between an open-minded, culturally and politically interested mathematician and a media and cultural theorist with a strong affection for media technologies is not only a possible, but a mandatory venture. As soon as one finds a mutual attractor which can serve as an operative bridging of Charles P. Snow's (in)famous gap between the humanities and the natural sciences into two separate academic cultures, the creative potential that Snow mentions in his text – and against all his emphasis on the separating factors – effectively can be gained. A collaboration under such a mutual gravitational concept then oftentimes generates far more intriguing questions and insights as the respective knee jerks of specialized disciplinary discourses. *Neighborhood Technologies* serves as such an attractor, and it thus can also be read as a concept that initiates novel academic neighborhoods.

3.

It is precisely the aforementioned three-dimensional layering of neighborhoods as object, conceptual principle, and media technology that serves as a heuristic guideline for this edition. *Section I: Neighborhood Epistemologies*, seeks as a first step to particularize the term *Neighborhood Technologies*. Mathematician Tobias Harks (Maastricht) and Computer Scientist Martin Hoefer (Aachen) provide an overview of current concepts, notions and definitions of neighborhoods in mathematics and computer science. Media theorist and historian Sebastian Vehlken (Lüneburg) examines the entwined media history of technical applications that implemented neighborhood concepts in order to generate models and explanatory modes for complex real-life phenomena. Furthermore, he depicts several examples of actual *Neighborhood Technologies* applications. In addition – as a case study on ad-hoc-networked traffic simulation – computer scientist Sándor Fekete (Braunschweig) explores how dynamic neighborhoods of computational agents contribute to a novel epistemology of computer simulation and how these provide an instructive media-technological environment for experimentation with the laws of physics.

The following sections, as a second step – and against the background of these systematic epistemological and historical foundations – expand the transdisciplinary discourse with regard to three fundamental dimensions of *Neighborhood Technologies*. In *Section II: Neighborhood Architectures*, Media and Cultural theorist Christina Vagt (Berlin) looks into the media history of architectural neighborhood planning models – with the works of Buckminster Fuller as its epicenter. In this course, Vagt's paper also reconstructs the becoming of neighborhoods as malleable, flexible and combinable units that hitherto were employed even to generate simulated world models. Complementary, architect Henriette Bier (Delft) inquires into contemporary digitally-driven architecture. With this she refers to architectural shapes that are not only designed and produced by digital means but are, actually, incorporating digital sensing-actuating mechanisms. These enable buildings to interact with their environment and users in real-time.

In *Section III: Neighborhood Societies*, mathematical sociologist and physicist Dirk Helbing (Zürich) presents a thorough re-conception of economic modeling. Helbing argues that networked decision-making and bottom-up self-regulation will be of growing importance in the face of the increasing complexity of socio-economic systems. In this light, he develops the concept of a *homo socialis* as a counter-model to the conventional *homo oeconomicus*. In an actual computer experiment, systems scientist Manfred Füllsack (Graz) investigates the emergence and probability of cooperation in repeated common-good-games under the influence of different network topologies. His multi-agent model interprets agents as regularly and precariously employed workers, and the results of his simulations indicate that the conditions supporting the *emergence* of cooperation might be less than optimal for the *maintenance* of cooperation and *vice versa*. Media historian Sebastian Giessmann (Siegen) traces the remarkable transformation of proximity and distance that took place in the history of credit cards and their mediated neighborhoods. He explores how this highly mobile medium of economic cooperation always depended on enabling infrastructural architectures, which have been – and still are – tied to local territories and sites of exchange.

Section IV: Neighborhood Interactions, discusses various forms of building dynamic neighborly relations in socio-technical and performative systems. Whilst media archaeologist Shintaro Miyazaki (Basel) examines the dimension of neighborhood sounding as a mode of organizing information transmission in early distributed communication networks, two articles explore the modes of interaction in swarm-like collectives: Sociologist Carolin Wiedemann (Hamburg) – with a case study on *Anonymous* and the *4chan* board – discusses the affective infrastructures of online swarms and asks about the effects of digital collectivity on concepts like emancipation and solidarity. Finally, theater and dance scholar Gabriele Brandstetter (Berlin) – by considering examples from contemporary dance and performance – in her essay contemplates the applied kinesthetic processes and impulses of control that serve as decisive principles for generating proximity or distance in collective bodies and that guarantee of cohesion in collective movements.

Thus, instead of trying to consistently represent the variety and large spectrum of distributed approaches to dynamic networks – a project which would have been doomed to failure from the beginning – with *Neighborhood Technologies* we installed a core attractor as shown by the contributors to this publication in different ways. Rather than bringing out a more-of-the-same book, we attempted to establish some new names and some extraordinary reading encounters for our respective scientific communities. The volume particularly seeks to provide a partly unfamiliar and novel playground that is meant to encourage further discussions and transdisciplinary thinking. With *Neighborhood Technologies* we want to offer and experiment with an updated passe-partout of concrete concepts and case studies of neighborhoods and their transformation from a mere *local coexistence* to an initiative force for multiple, complex and dynamic *relations, actions and behaviors*.

I. NEIGHBORHOOD EPISTEMOLOGIES

TOBIAS HARKS, MARTIN HOEFER

NEIGHBORHOODS IN MATHEMATICAL OPTIMIZATION AND ALGORITHMIC GAME THEORY

ABSTRACT

This article demonstrates how neighborhoods are used in mathematical optimization and algorithmic game theory in a dual way: as a conceptual building block for defining solution methods and as objects of study in the respective fields of research.

1 INTRODUCTION

In this article we describe how neighborhood-based concepts have influenced and shaped different areas in mathematical optimization and game theory. The main goal is to highlight the transdisciplinary use of neighborhoods in the respective fields of science. On the one hand, neighborhoods are used in mathematical optimization as a scientific method to solve difficult optimization problems. The problem is that for many real-world optimization problems the solution space is too large to be able to perform for exhaustive search. Additionally, there is strong evidence from complexity theory that such problems are *intractable* in the sense that there are instances for which an exact solution method would require too much time (exponential to problem size). The simple idea of *local search* is to start searching for a better solution in a predefined *neighborhood* of a given initial solution. Once a better solution in the neighborhood has been found, this new solution defines a new neighborhood and the search continues. If no better solution in a neighborhood is found the search terminates with a *local optimum*. While conceptually extremely simple, we will demonstrate that this search method has been applied successfully to many *hard* optimization problems and is still in use today.

On the other hand, we demonstrate that in algorithmic game theory, neighborhoods are used both as scientific objects and as scientific methods at the same time. One prime example of this dual use appears in Thomas C. Schelling's work on housing segregation in major US-American cities. He analyzes segregation dynamics in urban areas using a simple *local search* dynamic that can be roughly described as follows: two citizens are sampled at random, and they switch residential location whenever for both of them the number of neighbors of the same kind (Schelling uses skin color as the criterion) of the respective other location is preferred to their current home location. He observes the counter-intuitive phenomenon that this simple local dynamic creates macroscopic patterns leading to segregation despite the fact that the thresholds defining the notion of "location a is preferred over location b" for the citizens are set in a way that would not represent a racist attitude.

The present article is organized as follows. In Section 2 we give an overview of how neighborhoods are used in order to solve or approximately solve hard optimization problems. We also show how the concept of neighborhoods has been used in theoretical computer science to define a complexity class known as *polynomial local search* ($\mathcal{PLS}$). In Section 3, we give an account of the dual use of neighborhoods in algorithmic game theory. We demonstrate that natural improvement dynamics (where players in a game iteratively change their strategy) can be seen as local search problems. Based on this observation we relate the complexity of computing a Nash equilibrium to the complexity class $\mathcal{PLS}$. Finally, we show that in algorithmic game theory neighborhoods as objects naturally appear in segregation-, matching-, and network-creation games.

2 LOCAL SEARCH IN MATHEMATICAL OPTIMIZATION

Speaking of *local search* implies that some form of neighborhood must be involved. As described here local search constitutes a classical but still successful method for solving hard optimization problems. In order to have a precise definition of a *hard problem* we need to introduce some mathematical notation. We will, however, try to keep the exposition as simple as possible yet formal enough to convey how neighborhoods have found their way into mathematics.

2.1 Problems and Algorithms

We first discuss the meaning of a mathematical optimization problem and formalized ways to analyze solution methods or algorithms. A problem is defined by specifying all relevant parameters and precisely defining what an answer (or solution) should look like. If all parameters are explicitly specified we speak of an *instance* of the problem. The set of all problems (of the same type) is called a *problem class*. To give a concrete example, consider the famous traveling salesman problem (TSP), where the goal is to compute a tour of minimal length visiting all cities of a certain region. Here, a concrete instance of the TSP could consist of the region Germany with all of its cities. Distances between cities are taken from a map and thus all relevant parameters are specified. The problem class TSP consists of all possible TSP instances one might create.

An *algorithm* for a problem class is a procedure that computes a solution for any instance of the problem class. Here, we use the term procedure to highlight that an algorithm is usually implemented as a computer program. In this context, an algorithm is nothing more than a set of rules by which elementary computer operations are executed.

Algorithms can be *efficient* or less so. Again, consider the class TSP. Here, a naive algorithm would simply enumerate all possible tours of a given instance and select the one with minimum length. Clearly, for very small n ($n \leq 20$ and n denoting the number of cities) this might be feasible, but for larger n this is an intractable way since there are roughly n^n possible tours and, thus, even for $n = 100$ no computer can enumerateall tours in a reasonable time. To have some formalism to measure the performance of algorithms, we need to specify a computational model on which we can analyze the efficiency of algorithms. The most accepted model is the Turing machine model or random access model. In this model, we assume that an instance is *encoded* to be readable by a Turing machine. We do not go into detail here, so it suffices to think of *binary encoding*, where an instance I is represented on an array (whose entries are either zero or one) and the length of the array needed to represent I corresponds to the size of the input denoted by $\langle I \rangle$. For instance the binary encoding of a non negative number n, denoted by $\langle n \rangle$ requires an array of length $\log(n)$. An algorithm is now considered to be *efficient* or *polynomial* for a problem class $\mathcal{I}$, if it performs for every $I \in \mathcal{I}$ no more than $p(\langle I \rangle)$ basic operations, where $p : \mathbb{N} \to \mathbb{N}$ is a polynomial function. Thus, if problem sizes get larger, the increase in running time grows only polynomially (instead of exponentially).

2.2 NP-complete Problems

In theory and practice it is often very useful to know if a problem class is *easy* or *hard*. In order to categorize problems accordingly we need to introduce some more notation. We assume from now on that the problem at hand is a *decision problem* with two possible answers *"yes"* and *"no"*. Note that restricting to decision problems is not a significant restriction as we can always turn an optimization problem to a series of decision problems by asking " Is there a feasible solution with value z?". We denote the class of decision problems for which a polynomial algorithm exists by $\mathcal{P}$ and we say that these are the *easy* problems. More difficult is the definition of the hard problems. We denote the class $\mathcal{NP}$ as the class of decision problems where every instance for which the answer is "yes" has a proof that can be verified in polynomial time. This means that if someone gives us an instance of the problem and a certificate to the answer being "yes", we can check in polynomial whether or not it is correct. We give an example.

Example 2.1. *Integer factorization belongs to $\mathcal{NP}$. Given integers n and m, decide if there is an integer k with $1 < k < m$ such that k divides n. Clearly, this is a decision problem because the possible answers are "yes" or "no". If we get such n, m, k with $1 < k < m$ and somebody claims that k is a factor of n (the certificate) we can check the answer in polynomial time by performing the division $\frac{n}{k}$.*

Before we come to the definition of the class of hard problems we need the following notion of a polynomial reduction.

Definition 2.1. *Let P_1 and P_2 be two decision problems. P_1 is polynomially reducible to P_2 if and only if there exists a transformation $f: P_1 \rightarrow P_2$ such that for all $I \in P_1, f(I) \in P_2$ can be computed in polynomial time (polynomial in $\langle I \rangle$) and I is a yes-instance of P_1 if and only if $f(I)$ is a yes-instance of P_2.*

A problem P is $\mathcal{NP}$-complete if $P \in \mathcal{NP}$ and every problem Q in NP is polynomially reducible to P. Surprisingly based on the work of Cook[1], Karp[2] showed that the class of NP-complete problems is non-empty. Among other problems, for instance, the decision version of the above-mentioned TSP is in fact NP-complete.

The question of whether or not the two classes $\mathcal{P}$ and $\mathcal{NP}$ coincide is the famous $\mathcal{P} = \mathcal{NP}$ question which so far nobody has been able to settle, making this one of the great unsolved problems of mathematics. The Clay Mathematics Institute offers a reward of $ 1 million to anyone who has a formal proof that $\mathcal{P} = \mathcal{NP}$ or that $\mathcal{P} \neq \mathcal{NP}$.

2.3 Local Search

Given the empirical evidence that many problems that are $\mathcal{NP}$-complete have withstood exact solution approaches for large-scale instances, local search approaches are frequently used to compute high-quality solutions. The class of local search problems is defined as follows:

---

1 Stephen Cook, "The Complexity of Theorem Proving Procedures," *Proc. 3rd Annual ACM Sympos. Theory Comput.* (1971), p. 151–158.

2 Richard Karp, "Reducibility Among Combinatorial Problems," *Complexity of Computer Computations*, ed. R.E. Miller and J.W. Thatcher (New York: Plenum Press, 1972).

Definition 2.2. *A local search problem P is specified by a set of instances $\mathcal{I}$ and for every instance $I \in \mathcal{I}$, let F(I) denote the set of feasible solutions. We are given a cost function $c : F(I) \rightarrow \mathbb{Z}$ that maps every $S \in F(I)$ to the cost value $c(S)$. For a solution $S \in F(I)$, we let $N(S, I) \subseteq F(I)$ denote the* neighborhood *of S. Thus, every $S' \in N(S, I)$ is in some sense close to S.*

We can associate a *transition graph* with an instance I of a local search problem P: Every solution $S \in F(I)$ corresponds to a node $v(S)$. There is a directed arc from $v(S_1)$ to $v(S_2)$ if and only if $S_2 \in N(S_1, I)$ and $c(S_2) < c(S_1)$. The sinks of this graph are the *local optima*. The simplest form of a local search algorithm is *iterative improvement* that follows a path in the transition graph. We will illustrate this method using the above-mentioned traveling salesman problem.

Example 2.2. *In the 2-OPT local search (originally proposed by Croes)*[3] *one iteratively deletes two edges, thus breaking the tour into two paths, and then reconnects those paths in the other possible way in case this gives a shorter tour. Hence, every possible tour T corresponds to a node in the transition graph and there is a directed edge between two nodes T_1 and T_2 if we can obtain a shorter tour T_2 by simply deleting two edges of T_1 and reconnecting as in Figure 1.*

2.4 The complexity class $\mathcal{PLS}$

Applying the local search paradigm a natural question to ask is: how difficult is it to actually compute a local optimum? This question lead in fact to the definition of the complexity class $\mathcal{PLS}$, defined as follows:

Definition 2.3. *A local search problem P belongs to $\mathcal{PLS}$, if the following polynomial time algorithms exist:*

1. *An algorithm A which computes an initial feasible solution $S \in F(I)$ for every instance $I \in \mathcal{I}$.*

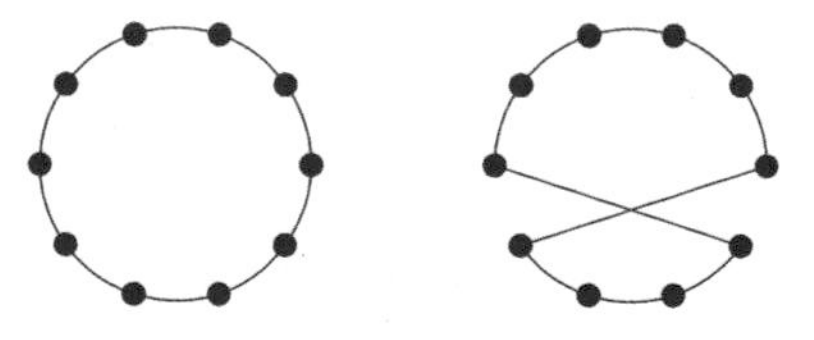

Fig. 1: A local improvement (2-Opt move): Original tour on the left and resulting tour on the right.

---

3 G. A. Croes, "A Method for Solving Traveling-Salesman Problems," *Operations Research* 6(6) (1958), p. 791–812.

2. *An algorithm B which computes the objective value $c(S)$ for every solution $S \in F(I)$ for every instance I.*

3. *An algorithm C, which determines for every instance I of P and every feasible solution $S \in F(I)$ whether S is locally optimal or not and finds a better solution in the neighborhood of S in the latter case i.e. it finds $S' \in N(S, I)$ with $c(S') < c(S)$.*

We illustrate the class $\mathcal{PLS}$ with the following example:

Example 2.3. *The problem* MAX CUT *with the* Flip-Neighborhood *is an example of a problem in $\mathcal{PLS}$. An instance I of* MAX CUT *consists of an undirected graph $G = (V, E)$ and arc weights $w : E \longrightarrow \mathbb{R}^+$. A feasible solution $S := (A, B) \in F(I)$ is a cut, i.e., a partition of $A \cup B = V$ of the node set V such that $A, B = \varnothing$. For a node set $U \subseteq V$ we define $\delta(U) := \{\{u, v\} \in E : u \in U, v \in V \setminus U\}$ and $w(\delta(U)) := \Sigma_{e \in \delta(U)} w(e)$. The goal is to find a cut (A, B) such that the weight of all arcs traversing the cut is maximal; that is, the objective of a cut $S = (A, B)$ is given by $c(S) := w(\delta(A))$. The Flip-Neighborhood $N(S, I)$ for a cut $S = (A, B)$ corresponds to all cuts (A', B') that are constructed from (A, B) by shifting a single node from A to B (or vice versa). Formally, $A' = A \cup \{x\}$ for some $x \in B$ or $B' = B \cup \{y\}$ for some $y \in A$. It can be seen that all conditions of Definition 2.3 are met.*

Similar to an $\mathcal{NP}$-reduction, we define a $\mathcal{PLS}$-reduction as follows:

Definition 2.4. Let $P_1 = (\mathcal{I}_1, F_1, c_1, N_1)$ *and* $P_2 = (\mathcal{I}_2, F_2, c_2, N_2)$ *be two problems in* $\mathcal{PLS}$. P_1 *is* $\mathcal{PLS}$-reducible *to* P_2*, if there are functions f and g that can be computed in polynomial time and (i) f maps every instance $I \in \mathcal{I}_1$ of P_1 to an instance $f(I) \in \mathcal{I}_2$ of P_2; (ii) g maps every tuple (S_2, I) with $S_2 \in F_2(f(I))$ to a solution $S_1 \in F_1(I)$; (iii) for all $I \in \mathcal{I}_1$: if S_2 is a local optimum of $f(I)$, then $g(S_2, I)$ is a local optimum of I.*

A problem P in $\mathcal{PLS}$ is $\mathcal{PLS}$-complete, if every problem of $\mathcal{PLS}$ can be reduced to P. For instance, MAX CUT is $\mathcal{PLS}$-complete.

3 NEIGHBORHOODS IN ALGORITHMIC GAME THEORY

Having established the notions of polynomial time algorithms and complexity classes like $\mathcal{NP}$ and $\mathcal{PLS}$, we are now ready to demonstrate the dual use of neighborhoods in the field of algorithmic game theory.

3.1 *Nash Equilibrium and the Improvement Graph*

One of the most fundamental concepts of game theory is the Nash equilibrium, an idea that has found wide-spread application in, e.g., mathematics, economics, biology or computer science. Nash equilibria are based on a local stability criterion, which gives rise to a neighborhood search method for computing them. While in general games this method might not be effective, there are prominent classes of games in which it succeeds.

Nash equilibria are stable outcomes in strategic games. Strategic games model competitive or cooperative scenarios, in which players interact with each other in a rational and strategic way. The interplay of strategy choices determines an outcome, and each player experiences a (possibly different) personal cost for each outcome. This creates incentives for players to pick strategies that optimize their cost.

More formally, in a finite *strategic game* there is a finite set of *players* $\mathcal{N}$. Each player $i \in \mathcal{N}$ has a finite set of actions or *strategies* $\mathcal{S}_i$. A *state* is a vector of strategies $S = (S_1, \dots, S_n)$ with $S_i \in \mathcal{S}_i$. Each player has a *personal cost* function c_i that maps each state S to a cost $c_i(S) \in \mathbb{R}$.

A classic example for a game is, e.g., the Prisoner's Dilemma. In this scenario two criminals committed a crime and are now interrogated separately by the police. Each prisoner has two strategies: (C)ooperate with the other prisoner and remain silent, or (D)efect and confess the crime. Depending on who confesses and who remains silent, each prisoner receives a different verdict, see Fig. 2. Although both prisoners keeping silent is by far the best outcome overall, there is a strong incentive for each of them to confess. Hence, if we assume rational choices and no coordination between players, then (D,D) is the only state from which no player unilaterally has an incentive to deviate.

	C	D
C	3,3	18,1
D	1,18	12,12

Fig. 2: A Prisoner's Dilemma game. Player 1 picks a row, player 2 picks a column. A state is a cell of the table. The first number is the cost of player 1, the second number the cost of player 2.

A *pure Nash equilibrium* is a a state from which no player has an incentive to unilaterally deviate given that all other players stick to their strategy choice. Formally, a state $S = (S_1, \dots, S_n)$ is a pure Nash equilibrium if for all players $i \in \mathcal{N}$, strategies $S'_i \in \mathcal{S}_i$ we have $c_i(S'_i, S_{-i}) \geq c_i(S)$, where $S_{-i} = (S_1, \dots, S_{i-1}, S_{i+1}, \dots, S_n)$. It is easy to verify that state (D,D) is a pure Nash equilibrium in the Prisoner's Dilemma. Nash equilibria capture the local preferences of a game, each player picks a *best response*, i.e., an optimal strategy choice against the current choices of other players. While the Prisoner's Dilemma has a unique Nash equilibrium, it is easy to image games with more than one Nash equilibrium, or even no Nash equilibrium at all.

The concept of Nash equilibrium suggests a natural local search procedure: iteratively select a player who can improve and let him switch to a better alternative. This represents a neighborhood search method over the set of states of a game. The neighborhood of a state S is given by all states S' that differ in the strategy choice of exactly one player i, and he strictly decreases his cost when moving from S to S'. More formally, the improvement graph of a game G_I consists of the set of states as vertices. There is a directed edge from state S to state S' if there is a player $i \in \mathcal{N}$ such that $S_i \neq S'_i$, $S_{-i} = S'_{-i}$, and $c_i(S') < c_i(S)$. The Nash equilibria of the game are exactly the vertices in the improvement graph without outgoing edges. The local search procedure described above constructs a sequence of profitable deviations of players, whereby it describes a path through the improvement graph.

In general games, there might be no Nash equilibrium (i.e., no vertex without outgoing edges exists in G_I), and then every such path must cycle eventually. In other games, Nash equilibria might exist but there is no path to reach them. Finally, although there are paths to one or more of the Nash equilibria, we might run into cycles by picking the wrong edges (the wrong sequence of players to deviate). While in general these problems can lead one to label the local search approach ineffective, there are prominent classes of games where all these problems can be avoided.

3.2 Congestion and Potential Games

The class of congestion or, more generally, potential games has found numerous applications when analyzing traffic networks or load balancing and scheduling problems. In a *congestion game*[4] there is a set N of players and a set R of *resources*. For each player $i \in N$ the strategy space $\mathcal{S}_i \subseteq 2^R$, i.e., each strategy $S_i \in \mathcal{S}_i$ is a subset of resources. For a resource $r \in R$ we denote by $n_r(S) = \{i \mid r \in S_i\}$ the *congestion* or *load* on resource r in state S, which is the number of players picking r in their strategy. Each resource r has a *delay* function $d_r(n_r(S)) \in \mathbb{Q}$ that gives the delay that the resource causes for the players that pick it. The cost for a player i is given by the sum of delays of chosen resources $c_i(S) = \Sigma_{r \in S_i} d_r(n_r(S))$.

Resources can be thought of as road segments in a traffic network or machines in a load-balancing scenario. Players strategically pick subsets of resources (routes through

---

4 Robert Rosenthal, "A Class of Games Possessing Pure-Strategy Nash Equilibria," *Int. J. Game Theory* 2 (1973), p. 65–67.

a network or machines) to minimize their total delay cost. For congestion games the improvement graph turns out to be acyclic – consider the following function Φ that assigns each state S a single numerical value:

$$\Phi(S) = \sum_{r \in R} \sum_{j=1}^{n_r(S)} d_r(j)$$

Suppose a single player i makes an improvement step, i.e., he changes his strategy from S_i to S'_i. Then we see that for states $S = (S_i, S_{-i})$ and $S' = (S'_i, S_{-i})$ we have

$$\Phi(S) - \Phi(S') = c_i(S) - c_i(S')$$

Thus, no matter which player deviates, Φ captures the change in his personal cost function. Put differently, there is a single function Φ for the states that simultaneously represents all personal cost functions of the players. It is now easy to see that the local search approach will terminate. In each improvement step, the potential strictly decreases. Thus, there cannot be cycles in the improvement graph. Since the game is finite, we reach a sink (i.e., a Nash equilibrium) in a finite number of steps.

The key property is the single function Φ that satisfies (1), i.e., it correctly mirrors all cost changes for unilateral deviations of all players in all states. A function Φ that has this property is called an *(exact) potential function*. Every strategic game that has a potential function Φ is called *(exact) potential game*. Observe that a potential game is just any game that enjoys the existence of some arbitrary potential function Φ that guarantees (1). It is not necessarily a congestion game defined via resources and delays. Clearly, every congestion game is a potential game. Perhaps surprisingly, the reverse is also true: Every potential game can, in fact, be interpreted as a congestion game.[5]

Consequently, we can apply the local search approach to find a Nash equilibrium in congestion and potential games. While we will find a Nash equilibrium eventually, it is unclear how many steps we need to take. After all, we are looking for a local optimum of the potential function within a neighborhood defined by deviations of single players. We can interpret the Max-Cut problem presented above as a potential game. In this game, each node $v \in V$ of the graph $G = (V, E)$ is a player and picks as strategy one of two sides. Each player receives as cost $c_v(S) = \Sigma_{\{u,v\} \in E: S_u = S_v} w(\{u, v\})$ the total value of incident edges to players on the same side. Hence, the incentive of each player is to maximize the total value of incident edges to players *on the other side*. If a player changes his strategy,

—

5 Dov Monderer and Lloyd Shapley, "Potential Games," *Games Econom. Behav.* 14 (1996), p. 1124–1143.

this corresponds to exactly a neighborhood move in the FLIP neighborhood described above. It is straightforward to verify that a potential function for this game can be given by $\Phi(S) = 2\sum_{\{u,v\}\in E: S_v = S_u} w(\{u, v\})$. Hence, computing a pure Nash equilibrium in a potential game is, in general, PLS-hard.

Given current complexity-theoretic beliefs, we do not expect to find an efficient algorithm to compute a pure Nash equilibrium in every potential (and congestion) game. There are, however, some prominent special cases for which the local search algorithm is known to converge quickly. For example, consider a congestion game in which every strategy of every player consists of exactly one resource, i.e. for all $i \in N$ and $S_i \in \mathcal{S}_i$ we have $|S_i| = 1$. Then every sequence of improvement moves has length at most $2n^2m$, where n is the number of players and m the number of resources.[6] This result can be generalized to congestion games with strategy spaces based on matroid structures, but it breaks for every more general class of games.[7] In fact, singleton and matroid structures play a prominent role for computation and convergence to Nash equilibria in many variants of congestion games, e.g., in bottleneck games where players consider the maximum delay as personal cost $c_i(S) = \max_{r\in S_i} d_r(n_r(S))$.[8]

3.3 Segregation Models and Dynamics

A classic example of studying neighborhoods in game theory are approaches to segregation, which originated in seminal work by Schelling.[9] Schelling tried to understand the effects that underlie segregation in urban housing and considered a very simple model. In the simplest form, he considers $2n$ players of two colors (n black and n white players) located in uniform distance along a ring. Every person has a utility for his location depending on his neighbors within a certain distance. In every round, we pick two arbitrary players and exchange their location if they both prefer to be in the location of the other person. Schelling observed experimentally that segregation evolves even if players have only a mild incentive to position themselves among peers of the same color. In par-

—

6 Samuel Ieong, Robert McGrew, Eugene Nudelman, Yoav Shoham, and Qixiang Sun, "Fast and Compact: A Simple Class of Congestion Games," *Proc. 20th Conf. Artificial Intelligence (AAAI)* (2005), p. 489–494.

7 Heiner Ackermann, Heiko Röglin and Berthold Vöcking, "On the Impact of Combinatorial Structure on Congestion Games," *J. ACM* 55(6) (2008).

8 Tobias Harks, Martin Hoefer, Max Klimm and Alexander Skopalik, "Computing Pure Nash and Strong Equilibria in Bottleneck Congestion Games," *Math. Prog.* 141(1–2) (2013), p. 193–215.

9 Thomas Schelling, "Dynamic Models of Segregation," *Journal of Mathematical Sociology* 1 (1971), p. 143–186; Thomas Schelling, *Micromotives and Macrobehavior* (New York: Norton, 1978).

ticular, even if only white players prefer just not to be in the minority in their neighborhood, segregation will typically evolve.

Neighborhood appears here on two different levels – on the one hand, the preference for each player is determined by the neighborhood of his position on the ring. On the other hand, the process that leads to segregation is a local search procedure over the set of states. It exchanges in each step exactly one randomly chosen pair of players that want to deviate. Many subsequent works have analyzed Schelling's findings from a more formal point of view, varying different aspects of the underlying topology or the dynamics of player deviation. We discuss two examples here.

Young[10] studies dynamics in which players are sometimes assumed to make mistakes. In each round, we pick a random pair of players from the ring. If they both want to deviate, we execute the switch. Otherwise, we assume players can make a *mistake*, i.e., with a small probability they switch even though this deteriorates their utility. By using mistakes, the underlying Markov chain of the game becomes ergodic and the dynamics never terminate. There exists a unique stationary distribution which describes the frequency of occurrence of each state in an infinite run of the dynamics. It turns out that complete segregation is the stochastically stable state, i.e., if we let the mistake probability tend to 0, the probability that we see a global segregation on the ring goes to 1. Hence, this result predicts that perfect global segregation evolves. This holds accordingly even for arbitrary numbers of black and white players and topologies other than the ring.[11]

A key to the latter result is the introduction of small mistakes to create nice analytical properties of the underlying Markov chain. The hope is that this small adjustment has no qualitative effect on the results, but it has recently been shown that, in fact, this is not true. Going back to the original model of Schelling, Brandt et al.[12] assume there are no mistakes, and in each round the randomly chosen pair of players switches only if this is in the interest of both. When each player considers a neighborhood of w players along the ring to make his choice, then starting from a uniformly random assignment they showed that the average block length (i.e., the length of a consecutive interval along the ring with players of the same color) is in the order of w. Instead of global segregation,

---

10 Hobart Peyton Young, *Individual Strategy and Social Structure: An Evolutionary Theory of Institutions* (New Jersey: Princeton University Press, 2001).

11 Junfu Zhang, "A Dynamic Model of Residential Segregation," *Journal of Mathematical Sociology* 28(3) (2004), p. 147–170.

12 Christina Brandt, Nicole Immorlica, Gautam Kamath and Robert Kleinberg, "An Analysis of One-Dimensional Schelling Segregation," *Proc. 44th Symp. Theory of Computing (STOC)* (2012), p. 789–804.

this result predicts only local segregation for Schelling's original approach. However, this property breaks even if we deviate slightly from the assumption of an equal number of black and white players, in which case the block lengths increase exponentially.[13]

3.4 *Contribution and Matching Games in Networks*

An area that recently started attracting increased attention is the understanding of fundamental coordination and network design problems in game-theoretic models with network locality. Usually, in these models players are actors embedded in a (social) network, which implies, e.g., limitations on information, restrictions for player utility, or locality of strategy spaces.

In a *contribution game* there is an undirected graph $G = (V, E)$, and each node $v \in V$ is a player. He is involved in several relationships with other players given by the incident edges. Player v has a *budget* $B_v > 0$ of effort (which is thought of as time or money) that he can invest in the relationships he is involved in. Formally, a strategy for each player v is a vector $(x_v^e)_{e:v\in e}$ that describes a non-negative numerical *contribution* for each incident edge $e = \{u, v\}$, where $x_v^e \geq 0$ and in total $\Sigma_{e:v\in e}\, x_v^e \leq B_v$. An edge e yields a *reward* $f_e(x_u^e, x_v^e) \geq 0$ that depends on the contributions of the incident players. Players invest their budgets strategically to maximize the sum of rewards of their incident edges. They can unilaterally decrease their contribution on an edge or pairwise increase both contributions on an edge (while possibly unilaterally decreasing it on other edges). This captures the intuition that successful investment in a relationship is a joint project, which might require unilateral dropping of others.

This model is analyzed by Anshelevich and Hoefer for a variety of different reward functions.[14] If the functions are convex (i.e., enjoy increasing marginal returns), players have an incentive to cluster all their effort on a single edge. In this case stable states exist, they essentially become pairings or *matchings* of players. In contrast, if reward functions are concave (with decreasing marginal returns), stable states have the property that players distribute their effort to balance the marginal returns as evenly as possible on their incident edges. In addition, the *price of anarchy* is studied, which is given by the ratio of the optimal total reward (if all players were investing their effort to maximize social welfare) over the total reward in the worst stable state. In many classes of games,

---

13 See George Barmpalias, Richard Elwes and Andy Lewis-Pye, "Tipping Points in Schelling Segregation," CoRR 1311.5934.

14 Elliot Anshelevich and Martin Hoefer, "Contribution Games in Networks," *Algorithmica* 63(1–2) (2012), p. 51–90.

the price of anarchy is upper bounded by 2, i.e. every stable state obtains at least half of the optimal total reward.

Contribution games are related to the classic stable matching problem in graphs. In this scenario, there are sets A and B of men and women embedded in an undirected, bipartite graph $G = (A \cup B, E)$ with $A \cap B = \emptyset$ and $E \subseteq A \times B$. Each person is a player and has a *preference order* $\succ_v$ over his neighbors in the graph. Players strive to match into pairs with preferred neighbors, where each player would rather be matched than unmatched. For a matching $M \subseteq E$ we denote by $M(v)$ the partner of player v, where $M(v) = v$ if v is unmatched in M. An edge $e = \{u, v\} \in E \setminus M$ is a *blocking pair* if $u \succ_v M(v)$ and $v \succ_u M(u)$. Intuitively, a blocking pair would give both players a strictly better partner than the one they get in M (if any). A blocking pair is problematic for stability – it is better for both u and v to come together and jointly create edge e. A matching without a blocking pair is called a *stable matching*.

Stable matchings have attracted immense research interest over the last decades (including recent Nobel memorial awards in economics) due to numerous applications in mathematics, computer science, and economics. A stable matching always exists and can be computed in polynomial time,[15] for arbitrary player neighborhoods emerging from a bipartite graph and preference orders of the players. Moreover, there are local search algorithms from which stable matchings can emerge. In particular, suppose we start with an arbitrary matching and iteratively resolve blocking pairs. In an *improvement step* we resolve a single blocking pair – we add the edge and for the incident players remove all edges to less preferred partners (if any). For every initial matching, there is a sequence of improvement steps that leads to a stable matching.[16] If we pick blocking pairs in an unfortunate way, however, we can also create cyclic sequences.[17] This is different for correlated games in which each edge $e \in E$ has a unique reward $f(e)$, and each player prefers incident edges with a higher reward. In this case, preferences are correlated via edge rewards. If we consider sequences of improvement steps for correlated games, it is easy to observe that upon creating an edge we only remove edges of smaller reward. This implies that the sorted vector of edge rewards can only increase lexicographically. Hence, a local search procedure that iteratively resolves blocking pairs

---

15 David Gale and Lloyd Shapley, "College Admissions and the Stability of Marriage," *Amer. Math. Monthly* 69(1) (1962), p. 9–15.

16 Alvin Roth and John Vande Vate, "Random Paths to Stability in Two-Sided Matching," *Econometrica*, 58(6) (1990), p. 1475–1480.

17 Donald Knuth, *Mariages stables et leurs relations avec d'autres problèmes combinatoires* (Montréal: Les Presses de l'Université de Montréal, 1976).

cannot cycle.[18] In this sense, there is an underlying lexicographic potential function. The local optima are exactly the stable matchings, and the problem of computing stable matchings belongs to PLS. However, the problem is not PLS-complete since a stable matching can be computed in polynomial time.

Neighborhood in matching and contribution games again manifests in two forms – the set of possible matching partners or joint contributors is limited to a static neighborhood. In addition, improvement dynamics create a neighborhood relation among states of the game. This provides local search methods to compute and converge to stable states. Hence, neighborhoods emerge both as an object of study and within methods for computation and analysis.

More recently, there have been several attempts to formalize an interaction between these two notions of neighborhood. In ordinary stable matching, each player has a fixed set of possible matching partners. However, when pairing up in social networks, players usually have a limited view on the population that evolves dynamically. A simple model to capture this idea is *locally stable matching*.[19] In addition to the graph $G = (A \cup B, E)$ of possible matching edges, we here have a network $N = (A \cup B, L)$ with a set of permanent links L. These links represent fixed relationships outside the matching interest (like, e.g., family members, or co-workers). Links can exist among any pair of players, i.e., $L \subseteq (A \cup B) \times (A \cup B)$. Here we assume each player has only a limited, dynamically changing snapshot of the population that he can access for matching. It is based on *triadic closure*, a standard concept in the study of social networks: If two individuals have a common friend, they are likely to meet. More formally, in a given matching $M \subseteq E$, two players u and v are *accessible* if $\{u, v\} \in E$ and they are within graph-theoretic distance of 2 in the network $N_M = (A \cup B, L \cup M)$. A *local blocking pair* is a blocking pair of accessible players (i.e., we now require unilateral improvement and access). A matching without local blocking pair is a *locally stable matching*.

In a locally stable matching, each player is embedded in a neighborhood of accessible partners that depends on who he is currently matched to. An improvement step resolving a local blocking pair depends on the current neighborhood of accessible partners. When resolving such a blocking pair, the dynamics move to a neighboring state, which creates a new neighborhood of accessible partners. In this sense, neighborhoods

---

18 David Abraham, Ariel Levavi, David Manlove and Gregg O'Malley, "The Stable Roommates Problem with Globally Ranked Pairs," *Internet Math.* 5(4) (2008), p. 493–515.

19 Esteban Arcaute and Sergei Vassilvitskii, "Social Networks and Stable Matchings in the Job Market," *Proc. 5th Intl. Workshop Internet & Network Economics* (WINE) (2009), p. 220–231.

of accessible partners depend on neighborhoods in the improvement dynamics and vice versa. Matching dynamics become a joint improvement and network exploration process. It might not come as a surprise that many nice computational features of ordinary stable matchings cannot be extended to locally stable matchings. Nevertheless, for correlated games we can still guarantee the existence of a lexicographic potential function and, more importantly, short improvement sequences to locally stable matchings.[20] This continues to hold also for a large variety of other variants of stable matching or more general coalition formation, in which neighborhoods can have different interpretations.[21] In addition, memory of players can drastically speed up local search and convergence dynamics for locally stable matchings.[22]

3.5 Network Creation Games

Matching games are an important elementary class of more general strategic network creation games. In such games, players are nodes in a network, and edges are created by strategic interaction. There are several streams of research in this broad area originating in game theory, sociology and computer science.

Jackson and Wolinsky proposed a strategic game in which rational players construct network edges.[23] Each player receives a reward, which is a function of the network. Edges can only be created if both players increase their reward in the resulting network. Edge deletion, however, can be conducted unilaterally. In a given state, either one pair of players can build a new edge, or a single player can destroy a single edge. If there is no profitable deviation of this kind, the network is called *pairwise stable*. Among other things, the authors studied two particular examples, the connections and the co-author model. In both models, players must pay a cost for their incident edges. They optimize a trade-off between edge cost and a benefit from their position in the network. In the connections model, each agent receives a direct benefit from every other agent that is weighted with a decaying factor depending on the distance in the network. In the

---

20 Martin Hoefer, "Local Matching Dynamics in Social Networks," *Inf. Comput* 222 (2013), p. 20–35.

21 Martin Hoefer and Lisa Wagner, "Matching Dynamics with Constraints," *Proc. 10th Intl. Conf. Web and Internet Economics (WINE)* (2014).

22 Hoefer, "Local Matching Dynamics in Social Networks;" Martin Hoefer and Lisa Wagner, "Locally Stable Marriage with Strict Preferences," *Proc. 40th Intl. Coll. Automata, Languages and Programming (ICALP)*, volume 2 (2013), p. 620–631.

23 Matthew Jackson and Asher Wolinsky, "A Strategic Model of Social and Economic Networks," *J. Econom. Theory* 71(1) (1996), p. 44–74.

co-author model, the benefit depends only on the direct neighbors and is inversely correlated with their degrees.

This strategic approach to network creation games has sparked a lot of interest in economics and game theory, and a wide variety of models (directed graphs, side payments, different benefit functions, etc.) and stability concepts (Nash equilibria, strong equilibria, etc.) have been considered over the last two decades.[24] Many of these works focus on existence, uniqueness or other structural properties of different equilibrium concepts but leave out the algorithmic perspective of equilibria and neighborhoods.

The first model within this class studied in computer science was a game where edge creation and deletion is unilateral.[25] Here each player can unilaterally create an arbitrary subset of incident edges or delete any subset of existing incident edges. A strategy $S_v \subseteq V \setminus \{v\}$ specifies the set of players to which v builds an edge. Edge creation here is unilateral – edge $\{u, v\}$ exists if either $u \in S_v$ or $v \in S_u$. A state S results in an undirected graph $G(S)$. The cost for player v in state S trades edge costs with sum of distances in the network, i.e., $c_v(S) = \alpha|S_v| + \Sigma_{\upsilon \neq V}\, dist_{G(S)}(u, v)$, where $\alpha > 0$ is a parameter of the game. The paper studies pure Nash equilibria of this game and provides several insights on existence, computational complexity, and the social cost of Nash equilibria.

A large variety of related models and adjustments (bilateral edge creation, non-uniform edge costs, side payments, strong equilibria, etc..) have been studied in recent years.[26] In particular, sum of distances is the basis for *closeness*, a classic node-based centrality measure in social networks[27] Other prominent centrality measures have been treated in the context of network creation games, such as eccentricity (maximum distance of v to any other node),[28] betweenness (number of shortest paths running over v),[29] or variants of closeness for disconnected networks.[30] The price of anarchy (social cost of

---

24 Matthew Jackson, *Social and Economic Networks* (New Jersey: Princeton University Press, 2008).

25 Alex Fabrikant, Ankur Luthera, Elitza Maneva, Christos Papadimitriou and Scott Shenker, "On a Network Creation Game," *Proc. 22nd Symp. Principles of Distrib. Comput.* (PODC) (2003), p. 347–351.

26 Éva Tardos and Tom Wexler, "Network Formation Games," *Algorithmic Game Theory*, chapter 19, ed. Noam Nisan, Éva Tardos, Tim Roughgarden, and Vijay Vazirani (Cambridge: Cambridge University Press, 2007).

27 Linton Freeman, "Centrality in Social Networks: Conceptual Clarification," *Social Networks* 1(3) (1979), p. 215–239.

28 Erik Demaine, Mohammad Taghi Hajiaghayi, Hamid Mahini and Morteza Zadimoghaddam, "The Price of Anarchy in Network Creation Games," *ACM Trans. Algorithms* 8(2) (2012), p. 13.

29 Xiaohui Bei, Wei Chen, Shang-Hua Teng, Jialin Zhang and Jiajie Zhu, "Bounded Budget Betweenness Centrality Game for Strategic Network Formations," *Theoret. Comput. Sci.* 412(52) (2011), p. 7147–7168.

30 Ulrik Brandes, Martin Hoefer and Bobo Nick, "Network Creation Games with Disconnected Equilibria," *Proc. 4th Intl. Workshop Internet & Network Economics* (WINE) (2008), p. 394–401.

worst Nash equilibrium over social cost of best state) has attracted a lot of interest in many of these models.

More recently, local improvement dynamics in network creation games have been studied, implying again two dimensions of neighborhood. Although in general these dynamics can cycle,[31] there are a variety of game classes in which iterative local search procedures converge to Nash equilibria (quickly).[32] Similar to locally stable matching, recent work started to intertwine the two dimensions of neighborhood with local and dynamically changing information about the network and restricted opportunities for edge creation.[33]

---

31 Brandes et al., "Network Creation Games with Disconnected Equilibria".

32 Pascal Lenzner, "On Selfish Network Creation," Ph.D. diss., Humboldt-Universität zu Berlin, 2014.

33 Davide Bilò, Luciano Gualà, Stefano Leucci and Guido Proietti, "Locality-Based Network Creation Games," *Proc. 26th Symp. Parallelism in Algorithms and Architectures (SPAA)* (2014), p. 277–286.

SEBASTIAN VEHLKEN

GHETTO BLASTS

MEDIA HISTORIES OF NEIGHBORHOOD TECHNOLOGIES BETWEEN SEGREGATION, COOPERATION, AND CRAZINESS

ABSTRACT

This article gives a media-historical overview of several seminal applications of *Neighborhood Technologies*, (1) in Cellular Automata (CA), (2) in Swarm Intelligence (SI), and (3) in Agent-based Modeling (ABM). It does by no way attempt to be exhaustive, but rather highlights some initial and seminal media-technological contributions towards a mindset which bears neighborhood principles in its core. The text thus centers around media technologies which are based upon the phenomenon that the specific topological settings in local neighborhoods give rise to interesting emergent global patterns which develop dynamically over time and which yield novel ways of generating problem solutions: Autonomy, emergence, and distributed functioning replace preprogramming, control, and centralization. *Neighborhood Technologies* thus can be understood on two levels: First, as specific spatial structures which initiate non-linear processes over time and whose results often cannot be determined in advance. As an effect, they provide media interfaces which visualize the interplay between local neighborhood interactions and global effects, e.g. in Cellular Automata (CA) where the spatial layout of the media technology enables dynamic processes and at the same time visualizes them as computer graphics. And secondly, they can be perceived as engines of transdisciplinary thinking, bridging fields like mathematical modeling, computer simulation and engineering. *Neighborhood Technologies* moderate between disciplines e.g. by implementing findings from biology in swarm-intelligent robot collectives whose behavior then is re-applied as an experimental setting for (con-)testing supposed biological factors of collective motion in animal swarms. The central thesis of this article is that *Neighborhood Technologies* by way of their foundation in neighborhood interaction make the notion of space utterly dynamic and transformable, and always intriguingly connected to functions of time. By their dynamic collective formation, *Neighborhood Technologies* provide decisive information about complex real-world phenomena.[1]

—

1 This article is based partly on some paragraphs of my Ph.D.thesis, published in German as Sebastian Vehlken, *Zootechnologien. Eine Mediengeschichte der Schwarmforschung* (Zürich-Berlin: diaphanes, 2012).

INTRODUCTION

When Babak Ghanadian and Cédric Trigoso founded their start-up *niriu* in Hamburg, Germany, in 2011, it was just another in a plethora of ideas for new social media platforms. In an alleged contemporary *zeitgeist* of a post-materialist and participatory culture of sharing[2] – where the acquired social relations and not the acquired property are said to be the new currency, and where former consumers, now able to engage in the development and design of products thanks to Web 2.0 and the like, transmute into prosumers – *niriu* tried to incorporate the concept of a local city neighborhood into the heart of their application. Or, as the marketing jargon in the *About Us* section of the website puts it:

> On niriu, you see right away what your neighbourhood has to offer – on the city map or on the list. Yoga teachers, hobby musicians or wine lovers: your neighbours are much more versatile than you think. Now it's time to finally get to know the people you pass in the street each day and to discover your neighbourhood! Borrowing a screw driver or taking part in a philosophical walk through the park – some things are only needed once a year, other things you'd like to try out but you don't know how or with whom. On niriu, you create offers or demands or you react to your neighbours' actions. Thus everybody profits from their neighbourhood's potential! Via niriu's profiles you're able to get a first impression of your neighbours. Thus you can find people with similar interests as you – if you'd like to meet them, you can make the first step online in order to get to know your neighbourhood in real life![3]

Regardless of how successful or not *niriu* has become since its founding phase, its intention to create an online network application that would support the revitalization of the somewhat aged idea of neighborly help and of lively neighborhoods in large cities seems to be noteworthy with regard to at least two aspects: On the one hand, it revalues

---

2 See e.g. Mirko Tobias Schaefer, *Bastard Culture! How User Participation Transforms Cultural Production* (Amsterdam: Amsterdam University Press, 2011); Axel Bruns, *Blogs, Wikipedia, Second Life, and Beyond: From Production to Produsage. Digital Formations* (New York: Peter Lang, 2008); Yochai Benkler, *The Wealth of Networks: How Social Production Transforms Markets and Freedom* (New Haven: Yale University Press, 2006); George Ritzer, Paul Dean and Nathan Jurgenson, "The Coming of Age of the Prosumer," *American Behavioral Scientist* 56/4 (2012), p. 379–398; Philip Kotler, "The Prosumer Movement. A New Challenge for Marketers," *Advances in Consumer Research*, 13 (1986), p. 510–513. See also the article of Dirk Helbing in this volume.

3 See the hompage of Niriu, https://niriu.com/niriu, October 10, 2012.

– similar to far more popular apps like e.g. *Foursquare* – the potential of global multimedia peer-to-peer communication as a form of social swarming that initializes dynamic social networks and interactions in public spaces.[4] *Niriu* thus defines itself as a kind of digital doorbell to next-door neighbors which circumvents the everyday obstructions of alleged anonymous neighborhoods of the contemporary metropolis. On the other hand, this example of a bottom-up city development initiative can systematically be seen as a re-engineering of Thomas Schelling's segregation models. Whilst *niriu* as an application engages with the adjustment of real neighborhoods and their issues, Schelling developed his modeling approach with regard to problematic neighborhoods, motivated to find novel explanatory modes for the relationship between housing neighborhoods and social networks. His models not only had a strong influence on the formal analysis of social networks, but must also be seen as a pioneering work for the present boom of an agent-based modeling and computer simulation paradigm (ABM). Schelling's work thus appeals to a media history of social simulations which tried to explore the local interactions *in* social networks and the development of those (and other) global scale effects which *Niriu* – *as* a social network – now tries to establish and intensify.

Between these two poles this article delves into some seminal examples which exemplarily substantialize our notion of *Neighborhood Technologies*. Furthermore, it draws on some genealogic developments between certain neighborhoods whose interaction principles concretize in specific media technologies whose social networking capacities likewise result in specific neighborhoods. Thereby, the text is not at all interested in completeness but in highlighting the complementary lines which define the intersection area of a knowledge *of* neighborhoods and a knowledge *by means of* neighborhoods.

The first part attends to a media history of Cellular Automata (CA) which from the 1950s onwards became a medium of mathematical modeling and a somehow playful approach to life-like, non-linear processes in systems of multiple elements, and which also put forth the often irreducible aspect of computer-graphical visualization that is mandatory for the understanding of such dynamics. It is the observation of system dynamics in space and time which lead to results that are closed to rigid analytical approaches. The second part deals with the zoo-technological and reciprocal developments of robotics, computer science and biology in the research area of swarm intelligence.[5] Inspired

—

4 See e.g. Geert Lovink and Miriam Rasch, eds., *Unlike Us Reader: Social Media Monopolies and Their Alternatives* (Amsterdam: Institute of Network Cultures, 2013).

5 See for a media-historical and -theoretical perspective e.g. Vehlken, Zootechnologies; Sebastian Vehlken "Zootechnologies. 'Swarming' as a Cultural Technique," *Theory, Culture and Society, special issue*

by the self-organization capacities of animal collectives like ants, fish or birds, robot engineers seek for distributed networking technologies which facilitate the development of autonomously moving robot collectives. And contrariwise projects like *RoboFish*[6] implement artificial technical agents as mini-robots in biological collectives, in order to explore their local interactions and collective behavior experimentally. And the third part focuses on some of essential constituents of the field of applications which from the 1990s onwards developed as the ABM paradigm. However, it will discuss ABM rather briefly as the following article of this section on *Neighborhood Epistemologies* by Sándor Fekete investigates ABM in the concrete case of traffic simulations.

1. CELLULAR AUTOMATA

Of Ghettos and Lily Ponds

It might just be a case of historical concurrence that Thomas Schelling's famous segregation model from 1969/1971 bears obvious similarities with CA, even if he assured his readers that he learnt about these techniques at a later date. John Conway's legendary *Game of Life*, after all, dates from 1968, and already in the 1940s Stanislaw Ulam and John von Neumann were working on a theory of self-replicating automata. Ulam's idea of implementing the theory of self-replicating automata on CA relieved von Neumann's initial model of the engineering problems resulting from a sort of mechanical primordial ooze.[7] I will come to this in the second paragraph of this chapter. Anyway: Seemingly unaware of such early computer simulation techniques Schelling manually played with an analogous set of local interaction rules which popularized Conway's bestiary of gliders, snakes, blinkers, beehives, or breeders around the same time.

Unlike the early advocates of CA, Schelling did not work with a self-reflective or computer-game-like approach, but he was interested in socio-economic phenomena, more concretely – as he would note later – in the connection of micromotives and emerg-

---

Cultural Techniques, ed. Geoffrey Winthrop-Young, Jussi Parikka and Ilinca Irascu (2012), p. 110–131; Jussi Parikka, *Insect Media. An Archaeology of Animals and Technology* (Minneapolis: University of Minnesota Press, 2011); Niels Werber, *Ameisengesellschaften. Eine Faszinationsgeschichte* (Frankfurt/M.: Fischer, 2014).

6 See the Biorobotics Lab of FU Berlin under http://biorobotics.mi.fu-berlin.de/wordpress/?works=robofish, October 25, 2014.

7 Stanislaw Ulam, "On Some Mathematical Problems Connected with Patterns of Growth of Figures," *Proceedings of the Symposium of Applied* Mathematics 14 (1962), p. 219–231; John von Neumann, *Collected Works*, ed. A.H. Taub (New York: Pergamon Press, 1963), p. 288–328.

ing macrobehaviors.[8] He occupied himself with the reasons for the emergence of ghettos in US cities, i.e. the differentiation into clear-cut racially identical neighborhoods. The common sense answer would imply a prevalent and profound racism. But could the phenomenon also result from other local motives, independent from ideological foundations? Schelling pursued this inquiry with the help of a schematic model. It functions according to a very simple scheme: He randomly distributed coins of two sorts on a checkerboard. If more than a defined number of unlike coins adjoin to a certain coin, this coin is again randomly placed on another unoccupied square. The result, mapped by way of one- and two-dimensional graphics on *paper tools*[9], is surprising: Even in scenarios with *mild tendencies of segregation* (that is, if the number of unlike coins in the 8-cell neighborhood must be high to make a coin move), on the macro level clearly separated accumulations of similar coins emerge very quickly. Schelling deduced that housing segregation thus not necessarily depends on racist ideologies, but could be effected by more neutral reasons such as seeking to not become a minority in a neighborhood. Nevertheless, and counter-intuitively, ghetto-like distribution patterns emerge from even such local preferences. Significant global patterns in dynamic neighborhood interactions thus can emerge even if these do not correlate explicitly or implicitly with the local preferences and objectives of the individual neighbors (Fig. 1).[10]

Where people settle, frogs are likely to dwell close by. And whilst Schelling mapped his dynamic segregation and aggregation processes in cities, the British evolutionary biologist William D. Hamilton played around with dynamic models of biological aggregation phenomena in the same year of 1971. Some time before he was retroactively celebrated by Richard Dawkins as one of the trailblazers of sociobiology and as a precursor of Edward O. Wilson,[11] and a decade before his fame culminated because of his game-theoretical and interdisciplinary collaboration with political scientist and *rational choice*-theorist Robert Axelrod on the evolutionary emergence of cooperation strategies,[12] he was eager to cast out the malign but lasting spirits of weakly defined

—

8 See Thomas C. Schelling, *Micromotives and Macrobehavior* (New York: Norton, 2006).

9 See Ursula Klein, *Experiments, Models, Paper Tools. Cultures of Organic Chemistry in the Nineteenth Century* (Stanford: Stanford University Press, 2003).

10 Thomas C. Schelling, "Dynamic Models of Segregation," *Journal of Mathematical Sociology* 1 (1971), p. 143–186.

11 See Richard Dawkins, *The Selfish Gene* (Oxford: Oxford University Press, 1976); Edward O. Wilson, *Sociobiology. The New Synthesis* (Cambridge: Harvard University Press, 1975).

12 Robert Axelrod and William D. Hamilton, "The Evolution of Cooperation," *Science* 211 (4489): 1390–1396 (1981).

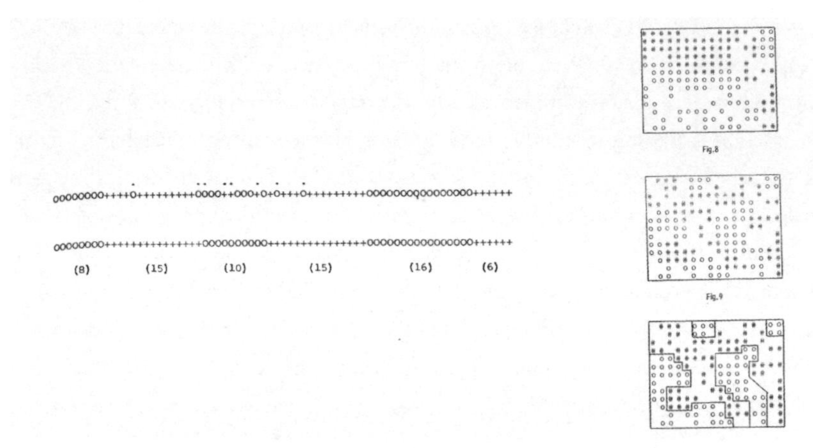

Fig. 1: Thomas C. Schelling: One-and two-dimensional Models of Segregation. Thomas C. Schelling, "Dynamic Models of Segregation," *Journal of Mathematical Sociology* 1 (1971), p. 143–186, 151 and 157.

so-called *social instincts* from behavioral biology.[13] Like Schelling, Hamilton abstracted from general social motives as catalysts of aggregation patterns in animal collectives. Instead, he proposed a geometrical model based on egoistic individual behavior – no signs of a cooperation theory at this stage of his career. In his model biological aggregations emerge solely from the individual actions of hypothetical frogs with regard to their spatial positioning in relation to their adjacent neighbors, and an external motivational factor, that is, a hypothetical predatory snake. No wonder that his paper starts like a mad fairytale:

> Imagine a circular lily pond. Imagine that the pond shelters a colony of frogs and a water snake. The snake preys on the frogs but only does so at a certain time of day – up to this time it sleeps on the bottom of the pond. Shortly before the snake is due to wake up all the frogs climb out onto the rim of the pond. This is because the snake prefers to catch frogs in the water. If it can't find any, however, it rears its head out of the water and surveys the disconsolate line sitting on the rim – it is supposed that fear of terrestrial predators prevents the frogs from going back from the rim – the snake surveys this line and snatches *the nearest one.*[14]

—

13 William D. Hamilton, "Geometry for the Selfish Herd," *Journal of Theoretical Biology* 31 (1971), p. 295–311.
14 Hamilton, "Geometry for the Selfish Herd," p. 295.

This set-up triggers quite a dynamic chain reaction. Given the ability of the hypothetical frogs to move unrestrictedly around the rim of the lily pond, they would seek to optimize their randomly taken relative positions. The danger to be the nearest one in relation to the snake can be reduced if a frog takes a position which is situated closely between two neighboring frogs. Put another way, the reduction of the individual *domain of danger*, that is, the sum of the distances to the next neighbors divided by 2 becomes the objective of each frog. This *domain of danger* certainly decreases if the next neighbors position themselves as close as possible. But as certain as this all the other frogs will also try to reduce their individual domains of danger. Or, as Hamilton notes: "[O]ne can imagine a confused toing-and-froing in which the desirable narrow gaps are as elusive as the croquet hoops in Alice's game in Wonderland" (Fig. 2).[15]

This model is played with one hundred hypothetical frogs, randomly spaced around the pond, and according to a simple algorithm:

> In each 'round' of jumping a frog stays put only if the 'gap' it occupies is smaller than both neighbouring gaps; otherwise it jumps into the smaller of these gaps, passing the neighbour's position by one-third of the gap-length. Note that at the termination of the experiment only the largest group is growing rapidly. The idea of this round pond and its circular rim is to study cover-seeking behaviour in an edgeless universe. No apology, therefore, need be made even for the rather ridiculous behaviour that tends to arise in the later stages of the model process, in which frogs supposedly fly right round the circular rim to "jump into" a gap on the other side of the aggregation. The model gives the hint which I wish to develop: that even when one starts with an edgeless group of animals, randomly or evenly spaced, the selfish avoidance of a predator can lead to aggregation.[16]

The relevance of Hamilton and Schelling's models for later developments of *Neighborhood Technologies* results from their approach which roots the emergence of global patterns in the autonomous decision making processes of locally interacting individuals. These interactions are mathematically described as topographical products (e.g., significant aggregations) of topological relations and are visualized as drawings of the dynamic neighborhood processes. An instructive insight from their approach is the fact that the respective global outcomes of collective processes can be totally independent from the individual objectives – like selfishness leading to aggregation. This interest in

---

15 Ibid., p. 296.
16 Ibid., p. 297.

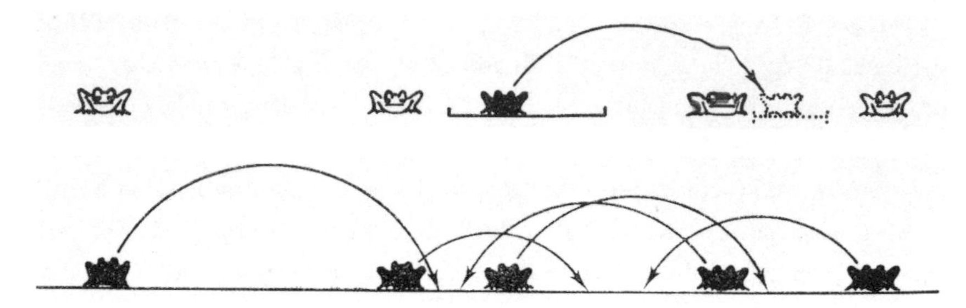

Fig. 2: William D. Hamilton: A confused toing-and-froing. William D. Hamilton, "Geometry for the Selfish Herd," *Journal of Theoretical Biology* 31 (1971), p. 295–311, 296.

the non-deducible effects of nonlinear interaction, even among very simple local *agents*, is investigated – as already mentioned – also in the field of computer science. Its media-technological implementation in CA will be the subject of the following part.

Of Zoos and Suburbs

Whilst economics and biology explore the micro- and macrodynamics of *living* collectives with checkerboards and frogs on paper around 1970, mathematics and the newly-developing computer science after World War II first spotlighted a different stage of life. At the end of the 1940s, John von Neumann started working on a general *Theory of Self-Reproducing Automata*, first presented at the *Hixon Symposium* in September 1948. Even if not explicitly defining his concept of an *automaton*, Neumann understands it as any system that processes information as a part of its self-regulatory mechanisms, as a system where stimulations effect specific processes which autonomously run according to a defined set of rules.[17] Leveling the ontological differences between computers and biological organisms,[18] von Neumann neither refers to mechanical parts or chemical or organic compounds as basic elements of his *bio-logics*, but *information*.[19]

Due to their systemic complexities, von Neumann compared the best computing machinery of his time with natural organisms. He discovered three fundamental boundary conditions for the construction of "really powerful computers": The size of

---

17 See Nancy Forbes, *Imitation of Life. How Biology is Inspiring Computing* (Cambridge: MIT Press, 2004), p. 26.

18 See Claus Pias, *Computer Spiel Welten* (München: sequenzia, 2002). Pias investigates the genealogy of CA back to military board games of the 19th century and the numerical meteorology of the early 20th century. The Hixon lecture was first published in 1951 and is also part of John von Neumann, Collected Works, p. 288–328.

19 See Steven Levy, *KL– Künstliches Leben aus dem Computer* (München: Droemer Knaur, 1993), p. 32.

10° segments of pool margin (degrees)

0 90 180 270 360

Position number																																				
1	2	3	3	3		6	1	3	3		4	1	3	5			1	3	4	5	2	9	4		5	6	3	5	4		1	2	2		4	3
2		5	2	2		8		3	2		6	1	1	7				2	4	7		11	2		7	5	2	5	5			3	2		4	4
3		6	1	2		8		3	1		8			9					4	9		11			9	4	1	7	5			2	2		4	4
4		6	1			9		3			9			8					4	11		10			10	3		8	6			1	2		4	5
5		7				9		2			9			8					5	12		8			12	1		9	6				3		4	5
6		7				9					9			8					5	14		7			13			9	5				5		3	6
7		8				7					9			8					5	16		5			15			9	3				6		3	6
8		9				5					9			9					4	18		3			17			9	1				6		5	5
9		10				4					8			10					3	20		1			19			8					6		7	4
10		12				3					7			11					2	22					20			6					6		9	2
11		13				2					6			12					1	22					22			4					7		9	2
12		14									6			13						22					24			3					7		9	2
13		14									5			13						22					26			2					8		8	2
14		15									3			13						22					28			1					9		7	2
15		16									1			13						22					30								10		6	2
16		17												12						22					32								9		6	2
17		17												11						22					33								8		7	2
18		18												9						22					35								6		8	2
19		19												7						22					37								5		9	1

FIG. 1. Gregarious behaviour of 100 frogs is shown in terms of the numbers found successively within 10° segments on the margin of the pool. The initial scatter (position 1) is random. Frogs jump simultaneously giving the series of positions shown. They pass neighbours' positions by one-third of the width of the gap. For further explanation, see text

Fig. 3: William D. Hamilton: Analysis of aggregation patterns. William D. Hamilton, "Geometry for the Selfish Herd," *Journal of Theoretical Biology* 31 (1971), p. 295–311, 297.

the building elements, their reliability, and the lack of a theory for the logical organization of complex computing systems. According to von Neumann, from the adequate organization of even unreliable components a reliability of the overall system could be produced that would exceed the product of the fault liability of the components.[20] "He felt," notes his editor Arthur W. Burks, "that there are qualitatively new principles involved in systems of great complexity." And von Neumann searches for these principles by investigating phenomena of self-reproduction, because "[i]t is also to be expected that because of the close relation of self-reproduction to self-repair, results on self-reproduction would help to solve the reliability problem."[21] Although he does not explicitly allude to collective dynamics, the preoccupation with robust and adaptive

—

20 Von Neumann describes these restrictions in more detail in John von Neumann, "Probabilistic Logics and the Synthesis of Reliable Organisms From Unreliable Components," *Collected Works*, vol. 5, p. 329–378.
21 John von Neumann, *Theory of Self-Reproducing Automata*, ed. Arhur W. Burks (Urbana/London: University of Illinois Press, 1966), p. 20.

features of a system involves a considerable conceptual affinity to the self-organizational capacities of dynamic networks. In both cases the adaptability to dynamically changing environmental conditions without a central control is decisive. But besides this conceptual link there is a media-technical relation that genealogically binds together the theory of automata and latterday ABM that can be perceived as a paradigmatic *Neighborhood Technology*. This link becomes only fully apparent after von Neumann's first hypothetical model – a mechanical model for self-reproduction later termed the *Kinematic Model* and supposedly inspired by von Neumann playing around with *Tinker Toys*[22] – was abandoned due to his collaboration with Stanislaw Ulam. Ulam proposed to rather use a different model environment which was unrestricted by the physical constraints of mechanical components, and far better suited for mathematical analysis.[23] His CA consisted of an infinite checkerboard as *biotope*, with every square of the grid potentially acting as a *cell* according to a program (a *State Transition Table*) effective for all the squares. Every cell is assigned information which defines its current state from a number of possible states. In each time-step of the system every cell updates its state, dependent on the number and states of neighboring cells, according to the transition table. The state of cell at time-step t+1 thus can be described as the function of the state of the cell and of all adjacent cells at t: "Every cell had become a little automaton and was able to interact with neighboring cells according to defined rules. [...] Thanks to Ulam's suggestion the science fiction of a tinkering robot in a sea of spare parts transformed into a mathematical formalism called 'cellular automaton'."[24]

—

22 This is executed in the following way: A construction part *A* (factory), produces an output X according to the instruction *b*(X). B functions as the copying part or duplicator and provides, on the basis of the input *b*, this *b* and an identical copy *b′* as output. C is the control unit which delivers the instruction *b*(X) to B. After the double-copying process of B, C delivers the first copy to *A* where the output X is produced, determined by the instruction *b*. Finally, C combines the remaining *b*(X) with the output of X produced by *A* and generates the output (X + *b* (X)) from the machine *A*+B+C. D is a particular instruction, enabling *A* to produce *A*+B+C. It is the self-description of the machine, D = *b*(*A*+B+C). The automaton *A*+B+C+D thus produces as an output precisely *A*+B+C+D without any sub-element reproducing individually. However, any sub-part is necessary for the self-replication of the whole machine. Self-organization thus is conceptualized as a system feature that characterizes the internal relation, the organization of sub-units. Herman Goldstine alluded to the anecdote that von Neumann used *Tinker Toys* in order to construct a three-dimensional model of his idea, see Herman H. Goldstine, *The Computer from Pascal to von Neumann* (Princeton: Princeton University Press, 1972), quoted in Robert Freitas Jr. and Ralph Merkle, *Kinematic Self-Replicating Machines* (Georgetown: Landes Bioscience, 2004). http://www.molecularassembler.com/KSRM/2.1.3.htm (last accessed September 22, 2014).

23 Compare Walter R. Stahl, "Self-Reproducing Automata," *Perspectives in Biology and Medicine* 8 (1965), p. 373–393, here 378.

24 Pias, *Computer Spiel Welten*, p. 259 (trans. Sebastian Vehlken).

Fig. 4: Von Neumann- and Moore-Neighborhoods on CA.

With this CA a self-reproducing artificial *organism* can be described in a mathematically exact way, in von Neumann's case mutating into a "monster" (Steven Levy) of 200.000 cells with 29 possible states each. The behavior and all state combinations were implemented on a square of 80 x 400 cells upon which all functions of the components A, B, and C were executed. Only the construction plan D was transferred into a one-dimensional *tail* of 150.000 cells. By reading and executing the information encoded in this *tail*, the CA was able to reproduce itself (and its reconstruction plan) as an identical duplicate. For von Neumann, this CA seems satisfactory enough as proof of a life-like form of self-organization. But more than this it becomes clear that CA – thanks to their conceptual basis – are suitable for the modeling of a multitude of dynamic systems:

> Compared to systems of differential equations, CA have the advantage that their simulations on digital computers do not produce round-off errors which can escalate especially in dynamic systems. Nonetheless, stochastic elements can be easily implemented in order to model disturbances. CA are characterized by dynamics in time and space. [...] Mathematically expressed this means that CA can be defined by: 1. the cell space, i.e. the size of its playing field, its dimension (line, plane, cube etc.), and its geometry (rectangle, hexagon etc.); 2. its boundary conditions, i.e. the behavior of cells which do not have enough neighbors; 3. the neighborhood, i.e. the radius of influence exerted on a cell (e.g. the 5-cell neighborhood of von Neumann, or the 9-cell neighborhood of Moore); 4. the number of possible states of a cell [...]; and 5. the rules that determine the change of states [...].[25]

CA can at best play off their epistemic potential if they are not only brought to paper as still images of process stages, but their dynamics can be computer-graphically displayed. Starting with the pioneering work of Burks and his research group at the University of Michigan this graphical approach has proved very fruitful. After the

—

25 Pias, *Computer Spiel Welten*, p. 257 (trans. Sebastian Vehlken).

possibility to computationally animate and interact with CA this media technology became a popular tool. Aside from the graphically supported *self-awareness* of CA, the relating anecdotes and the bestiary of aggregations of small squares of the *Game of Life*, CA became attractive for the application in other scientific disciplines because of their capability of exactly assigning particular characteristics to defined elements and the possibility to animate the interactions of these elements over time. As an effect, new scenarios could be observed and evaluated much faster.[26] And this – reminding the key word *lifelike behavior* – was not only instructive for biological research where CA were applied in behavioral studies, the computer simulation of animal collectives, but also in histology, neurology or in evolutionary and population biology. Also, in a socio-economic research context, CA-based computer simulation models were utilized. Even today, CA are a popular simulation technique in urban planning, e.g. in order to map the development of urban sprawl. The geographers Xiaojun Yang and C.P. Lo underline the advantages of CA in this field over other media technologies:

> Among all the documented dynamic models, those based on cellular automata (CA) are probably the most impressive in terms of their technological evolution in connection to urban applications. Cellular automata offer a framework for the exploration of complex adaptive systems because of CA's advantage including their flexibility, linkages they provide to complexity theory, connection of form with function and pattern with process, and their affinities with remotely sensed data and GIS.[27]

However, the authors also mention the conventional simplicity of cellular automata which has been considered one of its greatest weaknesses for representing real cities. But according to another seminal paper, a variety of research efforts of the 1990s have improved "the intricacies of cellular automata model construction, particularly in the modification and expansion of transition rules to include such notions as hierarchy, self-modification, probabilistic expressions, utility maximization, accessibility measures, exogenous links, inertia, and stochasticity."[28]

—

26 Compare G. Bard Ermentrout and Leah Edelstein-Keshet, "Cellular Automata Approaches to Biological Modeling," *Journal of Theoretical Biology* 160 (1993), p. 97–133.

27 Xiaojun Yang, C.P. Lo, "Modelling Urban Growth and Landscape Changes in the Atlanta Metropolitan Area," *International Journal for Geographical Information Science* 17/5 (2003), p. 463–488, here: p. 464.

28 Ibid., see also Paul M. Torrens and David O'Sullivan, "Cellular Automata and Urban Simulation: Where do we go from here," *Environment and Planning* B 28 (2001), p. 163–168.

Yang and Lo welcome the resulting *coming-of age* of CA as these – say the authors – "grow out of an earlier game-like simulator and evolve into a promising tool for urban growth prediction and forecasting,"[29] whilst Torrens and O'Sullivan also underline the necessity of combining these technological developments with further research in applied areas. These include explorations in spatial complexity, an undertaking which would call for the infusion of CA with concepts from urban theory, novel strategies for validating cellular urban models, as well as scenario design and simulation in relation to urban planning practices, e.g. by taking into consideration data from traffic simulation systems (Fig. 5).[30]

This example from urban planning brings to the surface a first media-historical element and its genealogy imperative for *Neighborhood Technologies*: With the models of Schelling and Hamilton and in the course of the development and the transdisciplinary application of CA entailing the pioneering work of von Neumann, Ulam and Conway, mathematical neighborhood models become alive in the artificial space of CA as dynamics in time. The non-linear and unpredictable interplay of neighborly micro-behaviors and the global systemic effects are implemented media-technologically and rendered feasible to analysis on a novel level of computer-graphically visualizations of discrete pattern emergence.

2. SWARM INTELLIGENCE

Of Dancing Drones and Robot Fishes

Even if it became popular in the context of the algorithmitization of the behavior of social insects, the birthplace of the term *Swarm Intelligence* is in robotics.[31] Even engineers are subject to discourse dynamics: When Gerardo Beni and Jing Wang gave a short presentation on *Cellular Robots* at a NATO robotics workshop in 1988, that is, on "groups

---

29 Xiaojun Yang, C.P. Lo, "Modelling Urban Growth," p. 465. See also Michael Batty and Yichun Xie, "From Cells to Cities," *Environment and Planning* B 21 (1994), p. 531–548; Helen Couclelis, "From Cellular Automata to Urban Models: New Principles for Model Development and Implementation," *Environment and Planning* B 24 (1997), p. 165–174; Yeqiao Wang and Xinsheng Zhang, "A Dynamic Modeling Approach to Simulating Socioeconomic Effects on Landscape Changes," *Ecological Modelling* 140 (2001), p. 141–162. For a comprehensive overview see Alison J. Heppenstall, Andrew T. Crooks, Linda M. See, and Michael Batty (eds.), *Agent-Based Models of Geographical Systems* (New York: Springer 2012).

30 See also the paper by Sándor Fekete in this volume.

31 Eric Bonabeau, Marco Dorigo and Guy Theraulaz, *Swarm Intelligence. From Natural to Artificial Systems* (New York: Oxford University Press, 1999).

Fig. 5: CA state of an urban sprawl simulation. Jean-Philippe Antoni, "Urban Sprawl Modeling: A methodological approach," 12th Colloque Européen de Géographie Théorique et Quantitative, St-Valéry-en-Caux, France, 7–11 September 2001, URL: http://cybergeo.revues.org/4188 (27th September 2014).

of robots that could work like cells of an organism to assemble more complex parts," commentators allegedly demanded a buzzword "to describe that sort of 'swarm'."[32] As an effect, Beni and Wang published their paper under the header *Swarm Intelligence in Cellular Robotic Systems*, coining a term which in the following years was employed in biological studies[33] and mathematical optimization problems[34] before gaining traction in the mainstream of robotics several years ago.[35] First, design approaches to distributed robot collectives were mainly inspired by research on social insects and relating computer simulation models (see the following chapter of this paper). But today, and in the course of developing *Unmanned Aerial Vehicles* (UAV) as drone collectives for military or civil use, the interaction modes of animal collectives operating in four dimensions, such as schools of fish or flocks of birds come also into focus.[36]

The basic interest is the question how complex global patterns of multiple individuals can emerge from simply structured, (mostly) identical, autonomously acting elements which interact only over short distances and are independent of a central controller or a central synchronizing clock.[37] The computer scientist Erol Sahin defines the field as follows: "Swarm robotics is the study of how a large number of relatively simple physically embodied agents can be designed such that a desired collective behavior emerges from the local interactions among agents and between the agents and the environment."[38] Such a concept is – at least theoretically – superior to centrally controlled and more

—

32 Gerardo Beni, "From Swarm Intelligence to Swarm Robotics," *Swarm Robotics*, ed. Erol Sahin and William M Spears (New York: Springer, 2005), p. 3–9, here: p. 3. Beni gives Alex Meystel the credit for bringing up the term in the discussion.

33 Compare Bonabeau, Dorigo and Theraulaz, *Swarm Intelligence*.

34 See also James Kennedy and Russell C. Eberhart, "Particle Swarm Optimization," *Proceedings of the IEEE International Conference on Neural Networks* (Piscataway: IEEE Service Center, 1995), p. 1942–1948.

35 See e.g. Alexis Drogoul et al., *Collective Robotics: First International Workshop. Proceedings, Paris July 4–5, 1998* (New York: Springer, 1998) Serge Kernbach, *Handbook of Collective Robotics* (Boca Raton: CRC Press, 2013).

36 See Joshua J. Corner and Gary B. Lamont, "Parallel Simulation of UAV Swarm Scenarios," *Proceedings of the 2004 Winter Simulation Conference*, ed. R. G. Ingalls, M. D. Rossetti, J. S. Smith and B. A. Peters (Piscataway: IEEE Press, 2004), p. 355–363, here: p. 355, referring to Bruce Clough, "UAV Swarming? So what are those swarms, what are the implications, and how do we handle them?," *Proceedings of 3rd Annual Conference on Future Unmanned Vehicles*, Air Force Research Laboratory, Control Automation (2003); Ferry Bachmann, Ruprecht Herbst, Robin Gebbers and Verena Hafner, "Micro UAV-Based Geo-Referenced Orthophoto Generation in VIS+NIR for Precision Agriculture," *International Archives of the Photogrammetry, Remote Sensing and Spatial Information Sciences*, vol. 40-1/W2 (2013), p. 11–16.

37 See Gerardo Beni, "Order by Disordered Action in Swarms," *Swarm Robotics*, ed. Sahin and Spears, p. 153–172, here: p. 153.

38 Erol Sahin, "Swarm Robotics. From Sources of Inspiration to Domains of Application," *Swarm Robotics*, ed. Sahin and Spears, p. 10–20, here: p. 12.

complex individual robots because of its greater robustness and flexibility as well as because of its scalability. Or, to put it shortly: "[U]sing swarms is the same as 'getting a bunch of small cheap dumb things to do the same job as an expensive smart thing'."[39] The large number of simple and like elements reduces the failure rate of critical functions and thus increases redundancy. Even if a number of robots fail, their functions can be replaced by other identical robots. This effect is complemented by the multiplication of the sensory capacities, "that is, distributed sensing by large numbers of individuals can increase the total signal-to-noise ratio of the system," which makes such collectives especially suited for search or observation tasks.[40] The capacity to automatically switch into various time-spatial patterns enables such robot collectives to develop modularized solutions for diverse situations by means of self-organization. They can adapt to unpredictable and random changes in their environments without being explicitly programmed to do so. And finally, they can be scaled to different sizes without affecting the functionality of the system.[41]

Essential for these capacities is the synchronization of the individual robots. In most cases, robot swarms update themselves *partially synchronously*:

> In fact, during an UC [updating circle, SV], any unit may update more than once; also it may update simultaneously with any number of other units; and, in general, the order of updating, the number of repeated updates, and the number/identity of units updating simultaneously are all events that occur at random during any UC. We call the swarm type of updating *Partial Random Synchronicity* (PRS).[42]

This equips each robot with a greater flexibility in relation to the adaptation on external factors since these can – due to the restricted sensing and interaction space of the robots – stimulate only a limited number of elements at a time. Beni calls this "order by disordered action."[43] The collective depends on spreading such influences from robot to robot and must be able to cope with time lags – a broad research area for stability analyses which prove e.g. that convergence and cohesion in swarming robots

—

39 Corner and Lamont, "Parallel Simulation of UAV Swarm Scenarios," p. 355.

40 Sahin, "Swarm Robotics," p. 12.

41 Ibid., p. 11.

42 Beni, "Order by Disordered Action in Swarms," p. 157.

43 Ibid., p. 153.

can be maintained in the presence of communication delays.[44] The specified number of neighborhood relations and the resulting spatial structure and morphology of mobile collectives shapes their synchronization processes. Time lags do not necessarily lead to less sustainable systems as the local transmission of information can moderate external influences throughout the collective. This dynamic stability thus not only seems to be a *time-critical* factor, but is also *space-critical*, dependent on the spatial ordering of a collective at a given time and on the topology of local interactions that determine the flow of information through the collective. The *formation* of the robot swarm produces *information*, and at the same time this information affects the formation of the collective.

An actual example is the COLLMOT (*Complex Structure and Dynamics of Collective Motion*) project at ELTE University of Budapest.[45] It explores the movement of swarms of drones, using neighborly control algorithms which can be found in biological swarms. By engineering actual autonomous quadrocopter collectives, the research group also endeavors to understand the essential characteristics of the emergent collective behavior which requires a thorough and realistic modeling of the robot and also of the environment in ABM before letting it take off as a physically built collective. The authors refer to the seminal model of collective motion of swarms proposed by computer engineer and graphic designer Craig Reynolds. In 1989 – at that time working as an animation designer for the Hollywood movie industry – he invented an individual-based flocking algorithm which not only resulted in the lifelike behavior of bat flocks in *Batman Returns*, but also opened up a whole new research area in biology in the following years. His animation model and its visualizations were quickly adopted by biologists who were interested in computer simulation approaches to animal collectives:[46]

> According to Reynolds, collective motion of various kinds of entities can be interpreted as a consequence of three simple principles: repulsion in short range to avoid collisions, a local interaction called alignment rule to align the velocity vectors of nearby units and preferably global positioning constraint to keep the flock together. These rules can be

—

44 Compare Yang Liu, Kevin M. Passino and Marios M. Polycarpou, "Stability Analysis of M-Dimensional Asynchronous Swarms With a Fixed Communication Topology," *IEEE Transactions on Automatic Control* 48/1 (2003), p. 76–95; compare Veysel Gazi and Kevin M. Passino, "Stability Analysis of Social Foraging Swarms," *IEEE Transactions on Systems, Man, and Cybernetics – Part B: Cybernetics* 34/1 (2004), p. 539–557.

45 See https://hal.elte.hu/flocking/wiki/public/en/projects/CollectiveMotionOfFlyingRobots (last accessed September 26, 2014).

46 For a more detailed perspective see Vehlken, *Zootechnologien*; see Craig W. Reynolds, "Flocks, Herds, and Schools: A Distributed Behavioral Model." *Computer Graphics* 21/4 (1987), p. 25–34.

> interpreted in mathematical form as an agent-based model, i.e., a (discrete or continuous) dynamical system that describes the time-evolution of the velocity of each unit individually. The simplest agent-based models of flocking describe the alignment rule as an explicit mathematical axiom: every unit aligns its velocity vector towards the average velocity vector of the units in its neighbourhood (including itself). It is possible to generalize this term by adding coupling of accelerations], preferred directions and adaptive decision-making schemes to extend the stability for higher velocities. In other (more specific) models, the alignment rule is a consequence of interaction forces or velocity terms based on over-damped dynamics. An important feature of the alignment rule terms in flocking models is their locality; units align their velocity towards the average velocity of other units within a limited range only. In flocks of autonomous robots, the communication between the robots usually has a finite range. In other words, the units can send messages (e.g., their positions and velocities) only to other nearby units. Another analogy between nature based flocking models and autonomous robotic systems is that both can be considered to be based on agents, i.e., autonomous units subject to some system-specific rules. In flocking models, the velocity vectors of the agents evolve individually through a dynamical system. In a group of autonomous flying robots, every robot has its own on-board computer and on-board sensors, thus the control of the dynamics is individual-based and decentralized.[47]

By programming and constructing UAV collectives, the COLLMOT group on the one hand certainly is interested in the increase of efficiency which a flock of drones can yield in comparison to single drones or other airborne technologies. As multiple units can cover an area much better while looking for a possibly moving target than a single drone, it can be used in search/rescue/hunt operations with onboard cameras/heatcams. Or it can be employed in agricultural monitoring, with a humming flock of drones preventing disasters befalling freshly-sprouting plants, measuring environmental conditions, assessing growth rate, or delivering nutrients or pesticides locally in small amounts. And in event surveillance, it could replace expensive cranes or helicopters and perform continuous surveillance or provide multiple viewpoints from the sky.[48]

But on the other hand, the engineered neighborhood technology of a UAV swarm also generates novel insights for the optimization and the assessment of ABM of col-

—

47 Tamás Viczek et. al., "Flocking Algorithm for Autonomous Flying Robots," *Bioinspiration & Biomimetics* 9/2 (2014), p. 1–11: 2. Http://iopscience.iop.org/1748-3190/9/2/025012/ (last accessed September 26, 2014).
48 See https://hal.elte.hu/flocking/wiki/public/en/projects/CollectiveMotionOfFlyingRobots.

lective behavior and through that generates a "reverse-bio-inspiration for biological research."[49] In this case, *Neighborhood Technologies* explicitly work as a bridge between scientific disciplines and between theory-building, modeling, and the construction of technical artifacts. And this process is a reciprocal one, as e.g. the COLLMOT group became again "inspired to search for additional factors which allow the very highly coherent motion of pigeon flocks, since our experiments suggest that a very short reaction time itself cannot account for the perfectly synchronized flight of many kinds of birds."[50]

Quite similarly, the *RoboFish* project of Freie University Berlin and of the Leibniz Institute of Freshwater Ecology make use of a reciprocal zoo-technological research perspective connecting robotics and biological studies. It develops a biomimetic fish school for the investigation of swarm intelligence.[51] In this case, the researchers control an artificial fish in a research aquarium populated by biological fish. By animating the RoboFish according to parameters of a variety of computer simulation models of fish schools, its influence on the behavior and the individual decisions of biological fish can be experimentally tested. And these findings then can be fed back to the CS models in order to make them more realistic. Accelerated by electromagnets under the aquarium floor, the effects of this *agent provocateur*[52] on the other schooling fishes can be tested. Jens Krause, one of the project leaders, e.g. identified certain *social thresholds*. Neighbors would assume that a fish with a more individualist swimming behavior possesses relevant information and follow it, whilst in larger schools, only a critical number of deviating individuals would be able to initiate a turnaround of the whole collective. Individual behavior – which always can also be a non-optimal movement – is moderated by the multiplicity of neighboring individuals with their respective local information, resulting in the optimized collective movement of the whole school with regard to external factors.[53] The absolute controllability of the RoboFish makes the definition of social thresholds for decision-making in animal collectives quantifiable and generates experimental data which contribute – in combination with CS models – to the biological knowledge of swarm intelligence.

---

49 Viczek et. al., "Flocking Algorithm for Autonomous Flying Robots," p. 10.

50 Ibid., p. 10.

51 See http://robofish.mi.fu-berlin.de/wordpress/?page—id=9 (last accessed September 26, 2014).

52 See Jakob Kneser, "Rückschau: Der Robofisch und die Schwarmintelligenz," *Das Erste – W wie Wissen*, broadcast of March 14, 2010, http://mediathek.daserste.de/daserste/servlet/content/ 3989916 (last accessed February 1, 2011).

53 Ibid.

Of Busy Ants and Crazy Particles

The collective capacity of social insects has fascinated naturalists in ancient times, the early behavioral biologists of the 18th and 19th centuries, and since the 1990s also a growing horde of computer scientists.[54] The latter became interested especially after Marco Dorigo et al. applied an optimization algorithm to the *Traveling Salesman Problem* (TSP) in the early 1990s which was inspired by food source allocation in ant colonies.[55] With reference to the communication structure of biological ants, they designed a system of individually foraging ant-like computational agents laying trails of simulated pheromones behind them which would evaporate over time. This would lead to two major effects: If one of the artificial ants would find a food source and then travel constantly from there to the nest and back, it would strengthen its pheromone trail, attracting other ants. And if several ants would find different ways to the food source, after a while the majority of the other ants would opt for the shortest way as pheromone trails on shorter routes contain more pheromones than those on longer routes. This capacity makes them an interesting modeling framework which has since become popular as *Ant Colony Optimization* (ACO). It involves a colony of cooperating individuals, an interaction by way of the individually sensed spatial environment, based on artificial pheromone trails for indirect (stigmeric) communication, a sequence of local moves in order to determine the shortest paths, and a probabilistic individual decision rule based on local information.[56]

ACO algorithms are suitable for solving problems that involve graph searching, especially when traditional approaches – e.g. dynamic programming – cannot be efficiently applied.[57] And over the last 15 years, they have been implemented e.g. for optimizing

---

54 See Leandro Nunes de Castro, *Fundamentals of Natural Computing: Basic Concepts, Algorithms, and Applications* (Boca Raton: CRC, 2007); Eva Johach, "Schwarm-Logiken. Genealogien sozialer Organisation in Insektengesellschaften," *Schwärme – Kollektive ohne Zentrum. Eine Wissensgeschichte zwischen Leben und Information*, ed. Eva Horn and Lucas M. Gisi (Bielefeld: transcript, 2009), p. 203–224; Niels Werber, *Ameisengesellschaften. Eine Faszinationsgeschichte* (Frankfurt/M.: Fischer, 2014); Parikka, *Insect Media*.

55 Alberto Colorni, Marco Dorigo and Vittorio Maniezzo, "Distributed Optimization by Ant Colonies," *Proceedings of ECAL91 - European Conference on Artificial Life, Paris, France* (Amsterdam: Elsevier Publishing, 1991), p. 134–142; Marco Dorigo, "Optimization, Learning and Natural Algorithms," Ph.D. diss., Politecnico di Milano, Italy, 1992; Marco Dorigo, Vittorio Maniezzo and Alberto Colorni, "Ant System: Optimization by a Colony of Cooperating Agents," *IEEE Transactions on Systems, Man, and Cybernetics - Part B*, 26(1) (1996), p. 29–41; Marco Dorigo, Mauro Birattari and Thomas Stützle, "Ant Colony Optimization: Artificial Ants as a Computational Intelligence Technique," *IEEE Computational Intelligence Magazine*, vol.1, numéro 4 (2006), p. 28–39.

56 See Marco Dorigo and Gianni Di Caro, "Ant Colony Optimization: A New Meta-Heuristic," *IEEE Congress on Evolutionary Computation - CEC '99*, ed. P. J. Angeline, Z. Michalewicz, M. Schoenauer, X. Yao and A. Zalzala (Piscataway, IEEE Press, 1999), p. 1470–1477.

57 See Nunes de Castro, *Fundamentals of Natural Computing*, p. 223.

telephone networks, vehicle routing, the coordination of manufacturing processes or the scheduling of working personnel. Optimization algorithms based on the self-organizational capacities of animal collectives thus can assist if optimization problems are at hand that have no analytical solution. One could say that the boundaries of calculability thus mark a movement towards biological principles.

Part of this field is also a stochastic optimization algorithm called *Particle Swarm Optimization* (PSO), introduced by mathematicians James Kennedy und Russell Eberhart in 1995.[58] Their model is based on the principles of how bird flocks find and circle around feeding grounds which are distributed in the environment. In the model, a so-called *cornfield vector* describes a target that motivates the exploration of the simulated particle swarm. The collective would – by way of distributing the individual sensory information about the perceived environment of each bird-oid agent to local neighbors – find this target faster than a systematic search of the complete simulation environment, advancing e.g. from the upper left corner to the one at the right bottom.

A swarm, after their definition, is "a population of interacting elements that is able to optimize some global objective through collaborative search of a space. Interactions that are relatively local (topologically) are often emphasized. There is a general stochastic (or chaotic) tendency in a swarm for individuals to move toward a center of mass in the population on critical dimensions, resulting in convergence on an optimum."[59] Hence, Kennedy and Eberhart develop their optimization algorithm by taking advantage of the dynamic relation between the individual perceptions and resulting movements and their influence on the collective movement of the collective on the one hand, and on the other by combining it with evolutionary algorithms in order to enable the simulation to learn:

> The method was discovered through simulation of a simplified social model. [... PSO] has roots in two main component methodologies. Perhaps more obvious are its ties to artificial life (A-life) in general, and to bird flocking, fish schooling, and swarming theory in particular. It is also related, however, to evolutionary computation, and has ties to both genetic algorithms and evolutionary programming.[60]

—

58 See Kennedy and Eberhart, "Particle Swarm Optimization;" compare Frank Heppner and Ulf Grenader, "A Stochastic Nonlinear Model for Coordinated Bird flocks," *The Ubiquity of Chaos*, ed. Saul Krasner (Washington: AAAS, 1990).

59 James Kennedy and Eberhart, *Swarm Intelligence* (San Francisco: Morgan Kauffman, 2001), p. 27.

60 Russell and Eberhart, "Particle Swarm Optimization," p. 1942.

PSO was first used to calculate the maxima and minima of non-linear functions, and afterwards also applied to multi-objective optimization problems, i.e. the optimization of a set of interdependent functions. In the latter case, the optimization of one function would conflict with that of the other functions, which makes it the object of PSO to moderate between these different demands – obviously in an optimal way. A possible example is a production process where all parameters that determine the production are interdependent, and where their combinations have an effect on the quantitative and qualitative outcomes of the process. To attain an optimal solution, theoretically all parameter combinations would have to be played through – and they increase exponentially, making this painstakingly interminable even with a relatively small number of parameters. Moreover, these parameters oftentimes are real and not integral numbers which makes the calculation of all possible cases simply impossible.[61]

PSO addresses these problems by examining the solution space of all possible parameter combinations by swarming, their bird-oid collectives being built along a simplified Reynolds model with the two basic parameters *Nearest-Neighbor Velocity* and *Craziness*.

At the beginning, the swarm particles are randomly distributed in the solution space, only defined by their position and their velocity. They are in contact with a defined maximum number of neighboring particles. The respective particle position at the same time designates a possible solution for the target function of the optimization problem. In iterative steps the algorithm now calculates the *personalBest* positions of the individual particles (the optimum is stored as a kind of personal memory) and the *neighborhoodBest* positions of the defined number of next neighbors. These are compared and evaluated, and in relation to (1) the best position, (2) the respective individual distances to this position, and (3) the former particle velocity, the direction and velocity of each particle is updated for the next iterative step. This would actuate a convergence of the local neighborhood towards the best position of the former time step.

However, the *Craziness* parameter is imperative for the later outcome, guaranteeing life-like swarming dynamics effected by incomplete information or external factors on the swarming individuals. *Craziness* simulates such disturbances by randomly interfering with the updated direction and velocity of a number of particles in each time step.

These variations prevent the particle swarm converging too fast around a certain position a.k.a. problem solution, as it could also be just a local optimum. And here the definition of the local neighborhood comes into play. If it is too large, the system shows

—

61 See Cai Ziegler, "Von Tieren lernen. Optimierungsprobleme lösen mit Schwarmintelligenz," c't 3 (2008), p. 188–191.

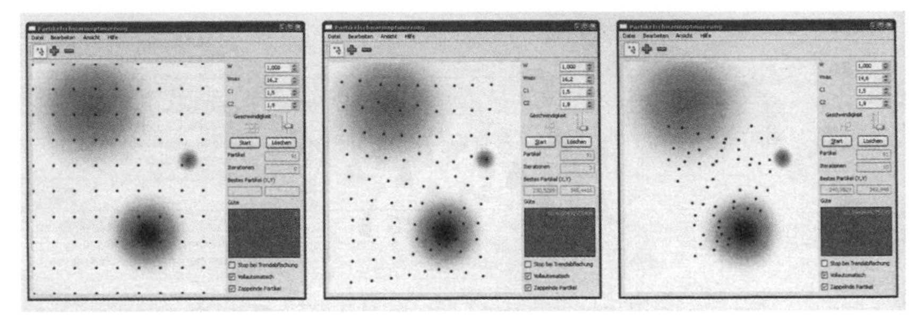

Fig. 6: Particle Swarm Optimization. Cai Ziegler, "Von Tieren lernen. Optimierungsprobleme lösen mit Schwarmintelligenz," *c't* 3 (2008), p. 188–191, here: p. 190.

the tendency to converge too early. If it is too small, the computation of the problem-solving process last longer. Similar to the swarming, hovering and the final convergence of biological bird flocks around a feeding ground, the particle swarms step by step aggregate at a certain position – the global maximum of the set of functions. The process of formation, the individual movements and neighborhood interactions indicate a mathematical solution. The dynamic spatial *formation* of the collective, based on local neighborhood communications, gives decisive *information* about the state of the environment.

3. AGENT-BASED MODELING

Of Kisses and Sugarscapes

Robert Axelrod knows his stuff. With many years of experience at the US Ministry of Defense and with *RAND Corporation*, he savors the fact that the meaning of the so-called *KISS principle* in US army jargon has a far less delicate meaning than one could assume. It just means: "Keep it simple, stupid."[62] According to Axelrod, of all possible instructions it is thus a military order which initiates the freedom of autonomous agents and ABM:

> The KISS Principle is vital [...]. When a surprising result occurs, it is very helpful to be confident that we can understand everything that went into the model. Although the topic being investigated may be complicated, the assumptions underlying the agent-based

—

62 Robert Axelrod, *The Complexity of Cooperation: Agent-Based Models of Competition and Collaboration* (Princeton: Princeton University Press, 1997), p. 3–4.

> model should be simple. The complexity of agent-based models should be in the simulations results, not in the assumptions of the model.[63]

Axelrod goes as far as describing agent-based modeling as

> a third way of doing science. Like deduction, it starts with a set of explicit assumptions. But unlike deduction, it does not prove theorems. Instead, an agent-based model generates simulated data that can be analyzed inductively. Unlike typical induction, however, the simulated data come from a set of rules rather than direct measurement of the real world. Whereas the purpose of induction is to find patterns in data and that of deduction is to find consequences of assumptions, the purpose of agent-based modeling is to aid intuition.[64]

The application of ABM thus transforms the modes of describing and of acknowledging dynamic systems. Joshua M. Epstein and Robert L. Axtell put this change of perspective as follows: "[ABM] may change the way we think about explanations [...]. What constitutes an explanation of an observed [...] phenomenon? Perhaps one day people will interpret the question, 'Can you explain it?' as asking 'Can you grow it?'"[65] In 1996, the authors published a computer program environment which refers to this task: Combining conceptual principles from Schelling's segregation models and Conway's *Game of Life*-CA, their modeling environment *Sugarscape* was designed as an interdisciplinary approach to explore complex dynamics in societies. Traditional sociological research would describe social processes in a more isolated way, only in some cases trying to aggregate them to "mega-models" (like the *Limits to Growth* report of the 1970s) with several shortcomings which attracted manifold critiques.[66] In contrast, their model would work as an integrative approach, bringing together societal subprocesses which were not easily decomposable.

The models consist of a number of artificial agents (inhabitants), a two-dimensional environment and the interaction rules governing the relations of the agents with each other and with the environment. Axtell and Epstein's original model is based on a 51x51 cell grid, where each cell can contain different amounts of sugar (or spice). In

---

63 Ibid., p. 5.
64 Ibid., p. 5.
65 Joshua M. Epstein and Robert L. Axtell, *Growing Artificial Societies: Social Science from the Bottom Up* (Cambridge: MIT Press, 1996).
66 William D. Nordhaus, "Lethal Model 2: The Limits to Growth Revisited," *Brookings Papers on Economic Activity* 2 (1992), p. 1–59.

every simulation step the individual agents search their local environment for the closest cell filled with sugar, move there and metabolize. As an effect of this process, they give rise to effects like polluting, dying, reproducing, inheriting resources, transferring information, trading or borrowing sugar, generating immunity or transmitting diseases – depending on the specific scenario and specific local rules defined at the set-up of the model.[67] Thus, say the authors, "the resulting artificial society unavoidably links demography, economics, cultural adaptation, genetic evolution, combat, environmental effects, and epidemiology. *Because the individual is multidimensional, so is the society.*"[68]

As physicist Eric Bonabeau notes, such ABM have become more and more popular in diverse fields of application since the 1990s – thanks to the availability of powerful computers and graphic chips which enable simulations with large agent populations and their interactions.[69] ABM, says Bonabeau, were superior to other methods of simulation for three reasons: First, they can reproduce emergent phenomena, second, they offer a *natural* system description, and third, they are flexible.[70] Unlike simulations that work through differential equations, agent-based models of simulation are based on the principle that *complex* global behavior can result from simple, locally-defined rules.[71] In Bonabeau's words:

> Individual behavior is nonlinear and can be characterized by thresholds, if then rules, or nonlinear coupling. Describing discontinuity in individual behavior is difficult with differential equations. [...] Agent interactions are heterogeneous and can generate network effects. Aggregate flow equations usually assume global homogeneous mixing, but the topology of the interaction network can lead to significant deviations from predicted aggregate behavior. Averages will not work. Aggregate differential equations tend to smooth out fluctuations, not ABM, which is important because under certain conditions, fluctuations can be amplified: the system is lineary stable but unstable to larger perturbations.[72]

—

67 See http://en.wikipedia.org/wiki/Sugarscape.

68 Epstein and Axtell, *Growing Artificial Societies*, p. 18.

69 See e.g. Roshan M D'Souza, Mikola Lysenko and Keyvan Rahmani, "SugarScape on Steroids: Simulating Over a Million Agents at Interactive Rates," *Proceedings of Agent 2007 conference* (Chicago, 2007); see also Sebastian Vehlken, "Epistemische Häufungen. Nicht-Dinge und Agentenbasierte Computersimulation," *Jenseits des Labors*, ed. Florian Hoof, Eva-Maria Jung, Ulrich Salaschek (Bielefeld: Transcript, 2011), p. 63–85.

70 Eric Bonabeau, "Agent-Based Modeling: Methods and Techniques for Simulating Human Systems," *PNAS* 99, Suppl. 3 (2002), p. 7280–7287: 7280.

71 See Forbes, *Imitation of Life*, p. 35.

72 Bonabeau, "Agent-based modelling," p. 7281.

In addition, it seemed more natural or obvious to model the behavior of units on the ground of local behavioral rules than through equations which stipulate the dynamics of density distribution on a global level. Also, agent-based simulation's flexibility made it easy to add further agents to an ABM and to adjust its parameters as well as the relations between the agents. An observation of the simulation system became possible on several levels, ranging from the system as a whole, to a couple of subordinated groups, down to the individual agent.[73]

However, we have to go back and ask: what exactly is an agent in the sense of ABM? Apparently, experts on the subject direct their attention to very different aspects. Bonabeau for instance considers every sort of independent component to be an agent, whether they are capable only of primitive reactions or of complex adaptive actions. In contrast, John L. Casti claims to apply the notion of the agent only to those components which are capable of adaptation and able to learn from environmental experiences in order to adjust their behavior when necessary; agents in that sense are only those components which consist of rules that enable them to change their rules.[74] In turn, Nicholas R. Jennings underlines the aspect of autonomy, meaning the ability of agents to take their own decisions and, therefore, to be defined as active instead of being only passively affected by systemic action.[75]

In defiance of such terminological differentiation, Macal and North list a number of qualities which are important from the pragmatic perspective of a model-maker or a *simulator*:

> An agent is identifiable, a discrete individual with a set of characteristics and rules governing its behaviors and decision-making capability. Agents are self-contained. The discreteness requirement implies that an agent has a boundary and one can easily determine whether something is part of an agent, is not part of an agent, or is a shared characteristic. An agent is situated, living in an environment with which it interacts along with other agents. Agents have protocols for interaction with other agents, such as for communication, and the capability to respond to the environment. Agents have the ability to recognize and distinguish the traits of other agents. An agent may be goal-directed, having goals

---

73 Ibid,, p. 7281.

74 See John L. Casti, *Would-be-Worlds. How Simulation is Changing the Frontiers of Science* (New York: John Wiley, 1997).

75 Nicholas R. Jennings, "On Agent-Based Software Engineering," *Artificial Intelligence* 117 (2000), p. 277–296.

> to achieve (not necessarily objectives to maximize) with respect to its behavior. This allows an agent to compare the outcome of its behavior relative to its goals. An agent is autonomous and self-directed. An agent can function independently in its environment and in its dealings with other agents, at least over a limited range of situations that are of interest. An agent is flexible, having the ability to learn and adapt its behaviors based on experience. This requires some form of memory. An agent may have rules that modify its rules of behavior.[76]

In the light of these basic features it is no surprise that the adaptive behavior of animal collectives – like Reynold's boid-model and the abovementioned ACO – are explicitly listed as examples and sources of inspiration for the ABM "mindset."[77] And the features lead to five decisive steps in the course of programming an ABM: First, all types of agents and other objects that are part of the simulation have to be defined in *classes* – together with their respective attributes. Second, the environment with its external factors and interaction potentials with the agents has to be modeled. Third, the agent methods are to be scripted, that is, the specific ways in which the agent attributes are subjected to updates as an effect of the interactions of the agent with other agents or with environmental factors. Fourth, the relational methods have to be defined that describe when, where and how the agents are capable of interacting with other agents during the simulation run. And fifth, such an agent model has to be implemented in software.

This set-up leads to a *procedural* production of knowledge: By varying the agent attributes and methods and observing the system behavior of the CS in interated simulation runs, the model is adjusted and modulated step-by-step. And the interesting thing with ABM is that they can be seen as a medium that integrates alternative approaches to engage with a problem (see also the articles by Sándor Fekete, Dirk Helbing and Manfred Füllsack in this volume):

> One may begin with a normative model in which agents attempt to optimize and use this model as a starting point for developing a simpler and more heuristic model of behavior. One may also begin with a behavioral model if applicable behavioral theory is available

---

76 Charles M. Macal and Michael J. North, "Tutorial on Agent-Based Modeling and Simulation. Part 2: How to Model with Agents," *Proceedings of the 2006 Winter Simulation Conference*, ed. L.F. Perrone et al., p. 73–83, here: p. 74.

77 Ibid., p. 75.

> [...], based on empirical studies. Alternatively, a number of formal logic frameworks have been developed in order to reason about agents [...].[78]

However, and somewhat provocatively, one could state that in ABM "performance beats theoretical accuracy":[79] the pragmatic *operationality* of the applications is more often than not more important than its exact *theoretical grounding*. And this performance is intrinsically linked to graphical visualizations of the non-linear and dynamic interplay of the multiple agents. Otherwise, the emerging model dynamics would remain untraceable in static lines of code or endless tables filled with quantified behavioral data, but lacking a dynamic time-spatial overall perspective on the system's developments in runtime.

Hence, with ABM the interdependencies of media and mathematics of dynamic networks become obvious in at least three ways: (1) The agent-based mindset animates mathematical models and thus generates novel perspectives on the emergence of global outcomes resulting from local interactions. (2) The ABM are capable of integrating diverse methodologies to address complex problems. And (3) with ABM the multidimensional effects of interaction processes can be investigated, thanks to the possible multidimensionality of agent attributes.

4. (NOT A) CONCLUSION

This article tried to give some seminal historical and contemporary examples of the media-technological genealogy of *Neighborhood Technologies*. Therefore, this part can be anything but a conclusion, since the paper merely wanted to present some landmarks which inspired us to take a transdisciplinary look at *Neighborhood Technologies*. Furthermore, the background of these historical examples might somehow serve as a framework and a reference plane for the diverse and fresh views on *Neighborhood Technologies* presented in the contributions to this volume. As it became apparent, the investigation of dynamic (social) networks has enormously profited from approaches that followed neighborly perspectives. From examples like Schelling's models of segregation onwards, it became also clear that only the coupling of mathematical modeling and analysis with

---

78 Ibid., p. 79.

79 Günter Küppers and Johannes Lenhard, "The Controversial Status of Simulations," *Proceedings of the 18th European Simulation Multiconference SCS Europe*, 2004.

media-technologies of computing or visualization would be suited for a multi-dimensional analysis of the emergent and unpredictable behavior of dynamic collectives. In *Neighborhood Technologies* thus the *investigation* of dynamic (social) networks and the *application* of such dynamics – described by mathematical models and programmed into the media technologies of an agent-based computer simulation mindset – are two sides of the same coin.

Furthermore, a media history of *Neighborhood Technologies* alludes to particular relations between spatial and temporal orderings. It is not only in cellular automata where the spatial layout of the media technology enables dynamic processes and at the same time visualizes them as computer graphics. Also in the abovementioned examples from swarm intelligence and ABM, the reciprocal interplay of topological and geometrical *formations* of the collectives and its internal *information* processing capabilities play a crucial role. This interplay is mediated by the respective definitions and sizes of local neighborhoods (how many and how distant members build a neighborhood etc.).

And not least, it became clear that *Neighborhood Technologies* can be perceived as media-technological engines of transdisciplinary thinking. They span fields like mathematical modeling, computer simulation, and engineering, e.g. by implementing findings from biology in swarm-intelligent robot collectives whose behavior then is re-applied as an experimental setting for (con-)testing supposed biological factors of collective motion in animal swarms. And it is precisely this transdisciplinary line of thought that our volume seeks to further pursue and encourage.

SÁNDOR P. FEKETE

NEIGHBORHOODS IN TRAFFIC

HOW COMPUTER SCIENCE CAN CHANGE THE LAWS OF PHYSICS

ABSTRACT

How can we explain and analyze the world around us? That is the key question any natural scientist has to answer when considering a complicated system. It is a standard approach in physics to reduce the underlying behavior to a set of rules that is as simple as possible, but still able to reproduce the observed behavior as well as possible. This has a close connection to neighborhood technologies: Very often, these rules are based on simple laws that govern the local interaction between neighboring basic components, and produce the global behavior in an emergent fashion. Once these laws are understood, one can try to exploit them for technological applications.

The possibilities of modern computer science have given a new perspective to this approach. In principle, we can program any set of physical laws for a complex system and then observe the resulting behavior – at least in simulations. What is more, we can go beyond figuring out absolute laws of physics, and *change* these laws, simply by programming different sets of rules for interaction.

In recent years, tremendous progress has been made in understanding the dynamics of vehicle traffic flow and traffic congestion by interpreting traffic as a multi-particle system. This helps to explain the onset and persistence of many undesired phenomena, e.g. traffic jams. It also reflects the apparent helplessness of drivers in traffic, who feel like passive particles that are pushed around by exterior forces; one of the crucial aspects is the inability to communicate and coordinate with other traffic participants.

The following article describes a distributed and self-regulating approach for the self-organization of a large system of many self-driven, mobile objects, i.e. cars in traffic. Based on short-distance communication between vehicles, and ideas from distributed algorithms, we consider reaction to specific traffic structures (e.g. traffic jams.) Building on current models from traffic physics, we were able to develop strategies that significantly improve the flow of congested traffic. Results include fuel savings of up to 40% for cars in stop-and-go traffic.

1. INTRODUCTION

1.1 Physics and Computer Science

A standard scientific method for *understanding* complicated situations is to analyze them in a top-down, hierarchical manner. This also works well for organizing a large variety of structures; that is why a similar approach has worked extremely well for employing computers in so many aspects of our life. On the other hand, our world has grown to be increasingly complex. The resulting challenges have become so demanding that it is impossible to ignore that a large variety of systems has a very different structure. The stability and effectiveness of our modern political, social and economic structures relies on the fact that they are based on decentralized, distributed and self-organizing mechanisms.[1]

Until not very long ago, scientific efforts for studying computing methodologies for decentralized complex systems had been very limited. Traditional computing systems are based on a centralized algorithmic paradigm: data is gathered, processed, and the result is administered by one central authority. Each of these aspects is subject to obstructions. On the other hand, "Living organisms [...] are to be treated as dynamic systems and contain all infrastructure necessary for their development, instead of depending on coupling to a separate thinking mind. We call this computing paradigm *organic computing* to emphasize both organic structure and complex, purposeful action."[2] The importance of organic computing has been motivated as follows: "The advantages of self-organizing systems are obvious: They behave more like intelligent assistants, rather than a rigidly programmed slave. They are flexible, robust against (partial) failure, and are able to self-optimize. The cost of design decreases, because not every variant has to be programmed in advance."[3]

Two important properties of organic computing systems are self-organization and emergence. We follow the definition of Tom De Wolf and Tom Holvoet[4]: "Self-organization is a dynamical and adaptive process where systems acquire and maintain structure themselves, without external control" and "[a] system exhibits emergence when there are coherent emergents at the macro-level that dynamically arise from the interactions

---

1 For a non-fiction bestseller see James Surowiecki, *The Wisdom of Crowds* (London: Doubleday, 2004).

2 Organic Computing Initiative. 2004. A novel computing paradigm. http://www. organic-computing.org.

3 Christian Müller-Schloer, Hartmut Schmeck and Theo Ungerer, *Organic Computing - A Paradigm Shift for Complex Systems* (Basel: Birkhäuser Verlag, 2011).

4 Tom De Wolf and Tom Holvoet, "Emergence versus Self-Organisation: Different Concepts but Promising when Combined," *Engineering Self-Organising Systems* 3464 (2005), p. 1–15.

between the parts at the micro-level. Such emergents are novel w.r.t. the individual parts of the system."[5]

In the following we will show how to combine aspects of distributed computing and communication (introduced in Section 1.2) with a new concept of algorithms and data structures, in order to deal with the challenge of influencing a very important complex system: traffic. Clearly, road traffic by itself (as introduced in Section 1.3) exhibits both properties postulated by de Wolf. Vehicles interact on the micro-level without external control by local interactions (self-organized) and due to these micro-level interactions, structures at the macro-level arise, such as traffic jams or convoys.

1.2 Distributed Communication

A critical aspect for influencing a complex system, no matter whether in a centralized or decentralized manner, is making use of distributed communication, which is becoming more and more feasible through the advance of modern technology. Today's applications for distributed systems rely on unicast or multicast communication. Prominent and well-known examples are Client-Server-Applications like web browsing. When mobile ad-hoc networks (MANETs) first came up, research concentrated on efficient routing to keep up the existing communication paradigm. By increasing the number of nodes and introducing dynamics in the network by means of node mobility or switching nodes on and off in a large multi-hop network, routing based on addresses became a hard challenge. In the meantime, systems have grown larger, to the point where they consist of thousands of cheap battery-powered devices that organize themselves. This has resulted in a new research field called *sensor networks*. Limitations of bandwidth, computing power and storage in sensor networks drive the corresponding communication paradigm. While these systems need to be designed for frequent node failures, redundancy was introduced in the network, decreasing the importance of single nodes and addresses. The stringent separation of the lower communication layers has also become open for discussion, as it introduces computational and networking overhead especially when identification of nodes loses importance. A recent survey article by Coulson et al.[6] provides an overview and a large number of more specific references; for further details, the reader is referred to the abundance on specialized literature.

—

5 Ibid.

6 Geoff Coulson, Barry Porter, I. Chatzigiannakis, C. Koninis, S. Fischer, D. Pfisterer, D. Bimschas, Torsten Braun, Philipp Hurni, Markus Anwander, Gerald Wagenknecht, S.P. Fekete, A. Kröller, T. Baumgartner, "Flexible Experimentation in Wireless Sensor Networks", *Commun. ACM* 55(1) (2012), p. 82–90.

1.3 *Traffic*

Traffic is one of the most influential phenomena of civilization. On a small scale, it affects the daily life of billions of individuals; on a larger scale, it determines the operating conditions of all industrialized economies; and on the global scale, traffic has a tremendous impact on the living conditions on our planet, both in a positive and in a negative way.

All this makes traffic one of the most important complex systems of our modern world. It has several levels of complexity, reaching from individual actions of the drivers, over local phenomena like density fluctuations and traffic jams, traffic participants' choice of transport mode and time, regional and temporal traffic patterns, all the way up to long-range traffic development and regulation. Clearly, this tremendous range of complexity also comprises different time scales as well as feedback effects, making traffic a particularly crucial yet difficult area, especially for critical strategic political decisions.

Over the years, tremendous progress has been made in understanding the dynamics of traffic flow and traffic congestion. Arguably the most significant contribution has come from physics, interpreting traffic as a multi-particle system. As described in further detail in Section 2.1, these models explain how the complexity of traffic emerges from the self-organization of individuals that follow simple rules. They also meet the popular appeal of systems of interacting, autonomous agents as problem-solving devices and models of complex behavior. In short, the *decentralized* view, which goes beyond attempts at centralized simulation and control, has improved our understanding of traffic.

1.4 *Combining Communication and Mobility*

Modern hardware has advanced to the point that it is technically possible to enable communication and coordination between traffic participants, in particular by vehicle-to-vehicle communication[7]. At this point, this is mostly seen as a mere technical gadget to facilitate driving, but the far-reaching, large-scale consequences have yet to be explored. Making use of the technical possibilities of communication and coordination should allow significant changes in the large-scale behavior of traffic as a complex network phenomenon. However, combining mobility and communication for coordinated behavior does not only solve problems, it also creates new ones, as it is a challenge in itself to maintain the involved ad-hoc networks, as well as the related information that is independent of individual vehicles.

---

7 E.g., see U.S. Department of Transportation, Research and Innovative Technology Administration, http://www.its.dot.gov/vii/

To this date and the best of our knowledge, the idea of combining the above aspects, i.e. self-organizing traffic by combining ad-hoc networks with distributed decentralized algorithms has received surprisingly little attention. This may be because it requires combining a number of different aspects, each of which has only been developed by itself in recent years: mobile ad-hoc networks, models for large systems of self-driven multi-particle systems, as well as algorithmic aspects of decentralized distributed computing, possibly with elements of game-theoretic interaction.

1.5 Our Work

Our basic idea is to develop a decentralized method for traffic analysis and control, based on a bottom-up, multilevel approach. Beyond the motivations described above, it should be stressed that an aspect of particular relevance is scalability:[8] while the computational effort for a centralized approach increases prohibitively with the number of vehicles, a decentralized method relies on neighborhood interaction of constant size.

2. RELATED WORK ON TRAFFIC

2.1 Traffic and Telematics

As the interest in guiding and organizing traffic has grown over the years, the scientific interest in traffic as a research topic has developed dramatically, as the excellent survey *Traffic and related Self-driven Many-Particle Systems* by Dirk Helbing shows.[9] Obviously, research on traffic as a whole is an area far too wide for a brief description in this short overview; we focus on a particular strain of research that is particularly relevant for our proposed work, as it appears to be most suited for simulation and extension to decentralized, self-organizing systems of many vehicles.

It is remarkable that until the early 1990s, efforts for simulating traffic were based on complex multi-parameter models of individual vehicles, resulting in complex systems of differential equations, with the hope of extending those into simulations for traffic as a whole. Obvious deficiencies of this kind of approach are manifold: First, because the

—

8 André B. Bondi, "Characteristics of Scalability and their Impact on Performance". *Proceedings of the Second International Workshop on Software and Performance – WOSP 2000*. p. 195. doi:10.1145/350391.350432; Mark D. Hill, "What is Scalability?". *ACM SIGARCH Computer Architecture News* 18 (4) (1990), p. 18.

9 Dirk Helbing, "Traffic and Related Self-Driven Many-Particle Systems," *Reviews of Modern Physics* 73 (2001), p. 1067–1141.

behavior of even just one vehicle is guided by all sorts of factors influencing a driver, the hope for a closed and full description appears hopeless. Second, determining the necessary data for setting up a simulation for a relevant scenario is virtually impossible. And third, running such a simulation quickly hits a wall; even with today's computing power, simulating a traffic jam with a few thousand individual vehicles based on such a model is far beyond reach.

A breakthrough was reached when physicists started to use a different kind of approach. Instead of modeling vehicles with ever-increasing numbers of hidden parameters, try to consider them as systems of many particles, each governed by a very basic set of rules. As Kai Nagel and Michael Schreckenberg showed, even a simple model based on cellular automata can produce fractal-like structures of spontaneous traffic jams, i.e. complex, self-organizing phenomena.[10] Over the years, these models[11] were generalized to two-lane highway traffic,[12] extended for simulating commuter traffic in a large city,[13] and have grown[14] to the point of being used for real-time traffic forecasts for the 2250 km of public highways in the German federal state of North Rhine-Westphalia.[15]

A closely related line of research uses an approach that is even closer to particle physics; the paper *Still flowing: approaches to traffic flow and traffic jam modeling*[16] gives an excellent overview of models for traffic flow and traffic jams, with about 150 relevant references. Among many others, particularly remarkable is the approach by Stefan Krauß.

---

10 Kai Nagel, Michael Schreckenberg, "A Cellular Automation Model for Freeway Traffic," *Journal de Physique* I France 2 (1992), p. 2221–2229.

11 Kai Nagel, "High-Speed Simulation of Traffic Flow," (Ph.D. diss., Center for Parallel Computing, Universität zu Köln, Germany, 1995).

12 Marcus Rickert, Kai Nagel, Michael Schreckenberg, Andreas Latour, "Two-Lane Traffic Simulation on Cellular Automata," *Physica* A 231 (1996), p. 534–550.

13 Marcus Rickert, Kai Nagel, "Experiences with a Simplified Microsimulation for the Dallas/Fort Worth Area," *International Journal of Modern Physics* C 8 (1997), p. 133–153.

14 Kai Nagel, "Cellular Automata Models for Transportation Applications," *Proc. 5th Int. Conf. Cellular Automata for Research and Industry* Vol. 2493 (2002), p. 20–31.

15 Oliver Kaufmann, Kai Froese, Roland Chrobok, Joachim Wahle, Lutz Neubert, Michael Schreckenberg, "Online Simulation of the Freeway Network of NRW," *Traffic and Granular Flow '99*, ed. Dirk Helbing, Hans J. Herrmann, Michael Schreckenberg, and Dietrich. E. Wolf (Berlin: Springer, 2000), p. 351–356. Andreas Pottmeier, Sigurdur Hafstein, Roland Chrobok, Joachim Wahle, Michael Schreckenberg, "The Traffic State of the Autobahn Network of North Rhine-Westphalia: An Online Traffic Simulation," *Proc. 10th World Cong. and Exh. on Intell. Transp. Syst. and Serv.* Doc. Nr. 2377 (2003). See also www.autobahn.nrw.de and Kai Nagel, "Traffic Networks," *Handbook of Graphs and Networks – From the Genome to the Internet*, ed. Stefan Bornholdt and Heinz Georg Schuster (Berlin: Wiley-VCH, 2003), Chapter 11.

16 Kai Nagel, Peter Wagner, Richard Woesler, "Still Flowing: Approaches to Traffic Flow and Traffic Jam Modeling," *Operations Research* 51, 5 (2003), p. 681–710.

This model reproduces properties of phase transitions in traffic flow, focusing on the influence of parameters describing typical acceleration and deceleration capabilities of vehicles.[17] This is based on the assumption that the capabilities of drivers to communicate and coordinate are basically restricted to avoid collisions, which until now is frustratingly close to what drivers can do when stuck in dense traffic.

Parallel to the scientific developments described above, the interest in and the methods for obtaining accurate traffic data has continued to grow. The employment of induction loops and traffic cameras has been around for quite a while, but despite of enormous investments, e.g., 200 Mio. Euros by the German Federal Ministry for Transport, Building and Urban Affairs (BMVBW) for putting up systems for influencing traffic, the limits on tracking individual vehicles,[18] as well as following particular traffic substructures are obvious. Another development is the use of floating car data. By keeping track of the movements of a suitable subset of vehicles (e.g., taxis in Berlin city traffic), the hope is to get a more accurate overall image of traffic situations, both in time and space.[19] However, even this approach relies on the use of the central processor paradigm, and does not allow the use of ad-hoc networks for the active and direct interaction and coordination between vehicles.

2.2 Traffic as a Self-Organizing Organic System

The structure of traffic is a phenomenon that is self-organizing at several levels.[20] But even though the behavior of and the interaction between motorists has been observed for a long time, the possibilities arising from modern technology allowing direct and decentralized complex interaction between vehicles has hardly been studied. The only efforts we are aware of combine game theory with traffic simulations - for example, the symposium organized by traffic physicist Schreckenberg with game-theory Nobel

---

17 Stefan Krauß, "Microscopic Modeling of Traffic Flow: Investigation of Collision-Free Vehicle Dynamics," (Ph.D. diss., Center for Parallel Computing, Universität zu Köln, 1998).

18 Bundesministerium für Verkehr, Bau und Stadtentwicklung, 2002. Programm zur Verkehrsbeeinflussung auf Bundesautobahnen 2002–2007. http://www.bmvbs.de.

19 Birgit Kwella, Heiko Lehmann, "Floating Car Data Analysis of Urban Road Networks," *Proceedings Computer Aided Systems Theory – EUROCAST '99*, ed. Franz Pichler, Roberto Moreno-Díaz, Peter Kopacek, *Lecture Notes in Computer Science*, vol. 1798 (Vienna: Springer, 2000).

20 For a philosophical discussion of self-organization in multi-level models see also Carlos Gershenson, Francis Heylighen, "When Can We Call a System Self-Organizing?," *Advances in Artificial Life*, 7th European Conference, ECAL 2003 (September 14–17, 2003), ed. Wolfgang Banzhaf, Thomas Christaller, Peter Dittrich, Jan T. Kim, Jens Ziegler, *Lecture Notes in Computer Science*, vol. 2801 (2002), p. 606–614.

laureate Reinhard Selten.[21] However, neither makes use of mobile ad-hoc networks and distributed algorithms in large networks.

Several research projects focus on enhancing traffic flow by distributed methods. A model based on multi-agent systems is described in Ana L. C. Bazzan's and Robert Junges' *Congestion tolls as utility alignment between agent and system optimum.*[22] Depending on local congestion, a dynamic toll is charged to influence the drivers' routing decisions. Camurri et al.[23] propose a distributed approach using Co-Fields for routing vehicles around crowded areas on an urban grid-like road map. Techniques for detecting traffic anomalies are also developed in centralized systems like the *System for Automatic Monitoring of Traffic* proposed by Bandini et al. The system uses video cameras which are installed along highways to derive and monitor traffic events over time based on the composite view of the cameras.[24]

All these approaches offer methods that try to solve certain problems arising in the field of traffic management. However, they cannot overcome the dilemma of an individual driver: The downside of individual freedom and control is the seemingly hopeless task of controlling a complex system that is highly sensitive to small perturbations. Even centralized control can try to delay the onset of the resulting oscillations – but it cannot remove the inherent instability of the overall system.

2.3 Our Approach

We have pursued a distributed and self-regulated approach for the self-organization of a large system of many self-driven, mobile objects, i.e., cars in traffic. Based on methods for mobile ad-hoc networks using short-distance communication between vehicles, and ideas from distributed algorithms, we revise the local physical interaction in congested traffic.

For this purpose, we developed strategies for dealing with complex and changing traffic situations, such as the improvement of traffic congestion. The basic idea is to

---

21 Michael Schreckenberg, Reinhard Selten, *Human Behavior and Traffic Networks* (Berlin: Springer, 2004).

22 Ana L. C. Bazzan, Robert Junges, "Congestion Tolls as Utility Alignment between Agent and System Optimum," *Proceedings of the Fifth International Joint Conference on Autonomous Agents and Multiagent Systems* (2006), p. 126–128.

23 Marco Camurri, Marco Mamei, Franco Zambonelli, "Urban Traffic Control with Co-fields," *Proc. of the 3rd Int. Workshop on Environments for Multiagent Systems* (2006), p. 11–25.

24 Stefania Bandini, Davide Bogni, and Sara Manzoni, "Alarm Correlation in Traffic Monitoring and Control Systems: A Knowledge-Based Approach," *Proceedings of the 15th European Conference on Artificial Intelligence* (2002), p. 638–642.

change the way in which individual vehicles respond to each other, and thus overcome the helpless situation of drivers in congested traffic.

3 TECHNICAL DETAILS

3.1 *Traffic Simulation*

Our approach for improving the flow of traffic is based on the well-known car-following model of Krauß which derives position and velocity of a car from the gap to the predecessor and its velocity.[25] Collisions are avoided based on a safe velocity:

$$v_{safe} = v_{pred} + \frac{g - \tau v_{pred}}{\frac{v + v_{pred}}{2b}}.$$

Here, v_{pred} is the velocity of the leading vehicle, g the gap to it, τ the reaction time (usually set to 1 s), and b the maximum deceleration. v_{safe} was proven by Krauß to prevent any collisions. To compute the desired velocity for the next time step, maximum velocity and acceleration a must be considered, too; for a simulation time step Δt, this leads to:

$$v_{des} = \min\left[v_{safe}, v_{max}, v + a\Delta t\right]$$

Finally, actual velocities may be lower than the maximum possible; this is accounted for by subtracting a random value from the current velocity, where *rand* returns a uniformly distributed value between 0 and 1, and σ is a configuration parameter between 0 and 1 to set the amount of randomness, which is usually set to 1.

$$v_{next} = \max[0, rand(v_{des} - \sigma a\Delta t, v_{des})]$$

Krauß's model is also the basis of SUMO [Hertkorn et al. 2002]. SUMO, an acronym for *Simulator of Urban MObility*, is an open-source microscopic traffic simulator developed by the German Aerospace Center DLR. Despite its name it can also simulate highway traffic. SUMO is used by many different traffic researchers all over the world.[26] An

---

25 Krauß, "Microscopic Modeling of Traffic Flow".

26 See http://sourceforge.net/apps/mediawiki/sumo/index.php?title=Projects.

alternative is Helbing's *Intelligent Driver Model.*[27] The details are mathematically intricate, but our approach yields similar results.

3.2 Improving the Flow of Highway Traffic

According to previous research,[28] one of the major reasons for collapses of traffic flow on highways, known as traffic jams, is the wide velocity distribution of vehicles. This is well-known to any experienced driver: even in dense traffic, high velocities are possible, as long as all vehicles move at almost the same speed; however, once a random fluctuation has occurred, the average speed drops considerably, and the overall pattern of motion becomes dominated by stop-and-go waves.

Overcoming non-uniform motion is a highly non-trivial matter. It is tempting to strictly enforce a uniform speed that seems to work so well before a collapse occurs. However, making large convoys of vehicles move in lockstep pushes the overall system into a highly unstable state; as a result, the catastrophic consequences of even a small failure or inaccuracy imply tremendous technical, legal, and psychological risks.

We pursued a gentler and more self-regulating alternative, in which individual drivers are still responsible for avoiding collisions. Instead, our driving strategy tries to avoid excessive and unnecessary acceleration when it can be determined that deceleration is imminent. The overall objective is not only to conserve fuel, but also to homogenize the overall speed distribution; as it turns out, this does improve the average speed. One additional, but equally important requirement was not to depend on the participation of all drivers; instead, even a relatively small system penetration should lead to measurable individual benefits, which constitutes an incentive for using the system and comply with its recommendations.

After a large variety of different tests and simulations, we have developed a recommendation that is based on a convex combination of the desired velocity of the driver and the average velocity of the vehicles.

$$v_{recommended} = \rho \, v_{desired} + (1 - \rho) v_{average}$$

---

27 Martin Treiber, Dirk Helbing, "Microsimulations of Freeway Traffic Including Control Measures," *Automatisierungstechnik* 49 (2001), p. 478.

28 See also Treiber, "Microsimulations of Freeway Traffic Including Control Measures" and Martin Schönhof, Dirk Helbing, "Empirical Features of Congested Traffic State and Their Implications for Traffic Modeling," *Transportation Science* 41, 2 (2007), p. 1–32.

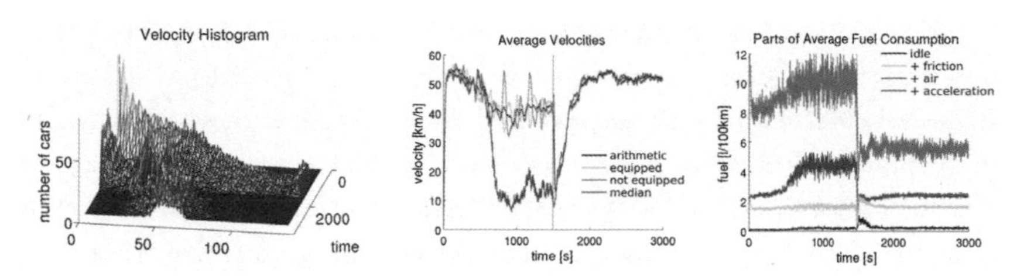

Fig. 1: The average speed increases significantly (top left), while the fuel consumption decreases by about 40% (top right). The key is a much tighter velocity distribution (center). This is achieved only through local interaction, without imposing a central speed limit.

Here ρ is a coefficient between zero and one. Setting $\rho = 1$ means not to recommend any different velocity, i.e. ignore the strategy. On the other hand, setting $\rho = 0$, i.e. ignoring the drivers' desired velocity completely is also counterproductive: whenever a vehicle ahead randomly slows down, vehicles following the strategy will also decelerate, implying that the average velocity converges to zero. Choosing the right compromise between these extremes works amazingly well. A patent[29] was granted in December of 2010. The graphics show typical simulation results, which result in about 40% fuel savings. They also show the key to this improvement: Based on simple local rules, the overall velocity distribution becomes much tighter, without artificially imposing or enforcing a global speed limit. Moreover, the rules provide individual benefits, even for single drivers, implying that the improvement is not dependent on the participation of all drivers. Finally, our approach is inherently *failsafe*: The strategy prevents drivers from unnecessarily accelerating, but drivers continue to be responsible for avoiding collisions (Fig. 1).

CONCLUSIONS

Interpreting vehicles in traffic as interacting physical particles has lead to considerable progress in understanding, analyzing, and even designing traffic scenarios. Our work is based on a computer science perspective: changing the interaction of individual components of a complex system can lead to a fundamentally different global behavior.

---

29 Sándor P. Fekete, Christopher Tessars, Christiane Schmidt, Axel Wegener, Stefan Fischer, Horst Hellbrück, *Verfahren und Vorrichtung zur Ermittlung einer Fahrstrategie*. Patentnummer DE 10 2008 047 143 B4 (2008).

We are convinced that this approach to complex systems will lead to progress in a wide range of other modern areas – ranging from large-scale (and even global) systems, all the way down to nano-scale systems that have been dubbed *programmable matter*. All this promises to yield exciting perspectives for science and technology, employing the concept of neighborhood technologies: The large-scale behavior of complex systems is often the result of simple local rules. Trying to obtain a different (ideally, more efficient) global behavior by dictating strict rules in order to exclude undesired emerging behavior is tempting, but very often futile; for example, outlawing sporadic braking for cars in stop-and-go traffic is no recipe for avoiding traffic jams. Instead, a more subtle, but also more powerful approach based on neighborhood technologies aims at modifying the underlying local rules that lead to the undesired emergent phenomena.

ACKNOWLEDGMENTS

This work was part of the DFG priority programme *Organic Computing*, supported by contracts FE 407/11-1, FE 407/11-2, FE 407/11-3. Different parts of this text are based on the publications [Fekete et al. 2010] and [Fekete et al. 2011]. We thank the colleagues involved, in particular, Björn Hendriks, Christopher Tessars, Axel Wegener, Stefan Fischer, and Horst Hellbrück.

II. NEIGHBORHOOD ARCHITECTURES

CHRISTINA VAGT

NEIGHBORHOOD DESIGN
BUCKMINSTER FULLER'S PLANNING TOOLS AND THE CITY

ABSTRACT

The article deals with the epistemological context of neighborhood design in the works of Richard Buckminster Fuller. Following the development of neighborhood planning models in 20th Century U.S. only briefly, the article wants to confront the concept of neighborhood technologies and social mathematics with a media history of environmental design and game theory, stress the central role of diagrams and graphics for this history and ask about its place within a broader history of gouvernmentality.

1. NEIGHBORHOOD UNIT

In the early 20th century, the *neighborhood* was reinvented as the smallest local unit in the social and political organization of a city. While Europe developed garden cities and *Neues Bauen*, the Chicago School of Sociology posed the question of the neighborhood from the perspective of human behavior and ecology.[1]

Dealing with problems of urban living such as traffic congestion, crowded living and working conditions and unfavorable environments, Clarence Perry developed the *neighborhood unit* as a formula for urban development in 1929. The diagrammatic planning model defines a neighborhood by a certain population size and a perimeter that restricts the maximum distance that the inhabitants have to cover to reach church, elementary school or shopping center (Fig. 1).

As Banerjee and Baer have pointed out in a broad study on urban neighborhoods in the U.S. from 1984, the debates on the neighborhood unit that became one of the most influential and criticized paradigms of urban design in the 20th century in the

---

1 See Robert E. Park, "The City. Suggestions for the Investigation of Human Behavior in the Urban Environment," *The City*, ed. Robert E. Park, Ernest Burgess, and Roderick D. McKenzie (Chicago: University of Chicago Press, 1984 [1925]), p. 1–46. See also Roderick D. McKenzie, "The Ecological Approach to the Study of Human Community," *ibid.*, p. 63–79.

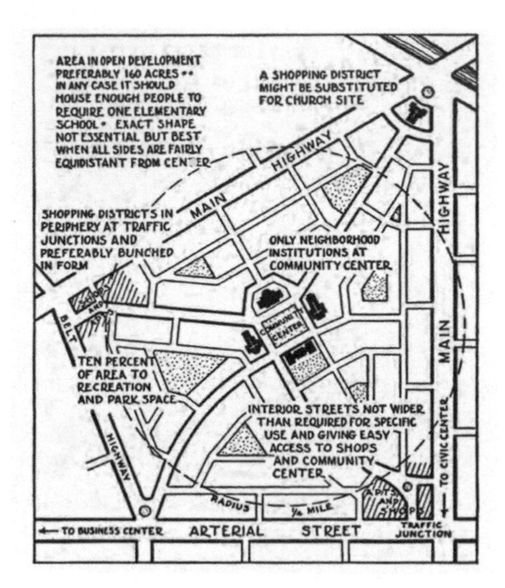

Fig. 1: Tridib Banerjee and William C. Baer, *Beyond the Neighborhood Unit. Residential Environments and Public Policy* (New York: Springer Science and Business Media 1984), p. 3

United States and other countries, were still viral in the late 1980s, and not only did they encapsulate the old debate between social and physical determinism, but they also point to the epistemological difference between sciences and design arts:

> "For scientists, the operative word is analysis – the separation of a whole into its component parts. For designers, the operative word is *synthesis* – the composition or combination of parts or elements so as to form a whole. The former understand the whole only after knowing about each of the parts in detail; the latter understand each part only after reference to the whole. This formulation is overstated to make a point. Science is not oblivious to synthesis; design is not unaware of analysis. But the guiding force is different in each case."[2]

In this paper, I want to follow the design epistemology of the neighborhood as synthesis and whole-part relation within urban planning up to the concept of a world designed by game theory in the 1960s and 1970s, to better understand the bio-political

---

2 See Tridib Banerjee and William C. Baer, *Beyond the Neighborhood Unit. Residential Environments and Public Policy* (New York: Springer Science and Business Media, 1984), p. 6.

impact of design within and beyond science.[3] By 1961, *neighborhood* had become a political controversy in and beyond urban planning, as Jane Jacobs resumes in her sociological informed monograph on *Death and Life of Great American Cities*: "Neighborhood is a word that has come to sound like a Valentine. As a sentimental concept, *neighborhood* is harmful to city planning. It leads to attempts at warping city life into imitations of town or suburban life."[4]

In the 1960s, schools, parks and clean housing no longer guaranteed a good life, and social and political control of cities had grown more complicated under rising social tensions and economical problems. Jacobs argues for "neighborhood planning units that are significantly defined only by their fabric and the life and intricate cross-use they generate, rather than by formalistic boundaries [...]. The difference is the difference between dealing with living complex organisms, capable of shaping their own destinies, and dealing with fixed and inert settlements, capable merely of custodial care (if that) of what has been bestowed upon them."

In need for new ways to think of cities, Jacobs turns to the history of scientific thought of cyberneticist Warren Weaver, whom she quotes at lenght. From Weaver, she adopts the idea of organisms as organized complexites. Therefore, cities are organic wholes of organized complexity, which have been mistaken by urban planning as problems of disorganized complexity.[5]

2. EKISTICS UNIT

Greek architect and urban planner Constantinos Doxiadis answered Jacobs' call for a new school of settlement planning in terms of life sciences (Fig. 2).

> "In our attempt to create a better environment for man we need to realize that people feel lonely if left alone in space and, therefore, tend to come closer together to form communities and an organized society. But these efforts may result in squeezing them so close together that they suffer. Thus the goal of the city is to bring people close enough

—

3 Compare the way *design* figures as production of the artificial and the new in opposition to classical knowledge production in the natural sciences: "The natural sciences are concerned with how things are. [...] Design, on the other hand, is concerned with how things ought to be, with devising artifacts to attain goals." (Herbert A. Simon, *The Sciences of the Artificial*, 3rd edition (Cambridge: MIT Press, 1996, p. 114–115).

4 Jane Jacobs, *The Death and Life of Great American Cities* (New York: Random House, 1961), p. 112.

5 Jacobs, *The Death and Life of Great American Cities*, p. 429–440.

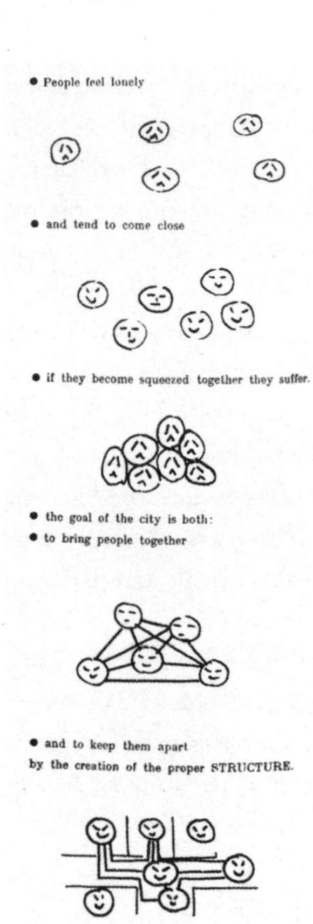

Fig. 2. C.A. Doxiadis, "A City for Human Development," EKISTICS (1969).

together to benefit from their contacts, but at the same time, to form a proper structure that can keep them sufficiently far apart, so that the exposure to and the danger from each other is minimized."[6]

In his article *A City for Human Development* for the 1968 June issue of the Ekistics magazine, Doxiadis problematizes the "inhuman city" as status quo and the need to re-examine the relation between humans and urban systems. According to Doxiadis the social behavior of humans is conditioned by their immediate environment (Fig. 3).

—

6 Constantinos Doxiadis, "A City for Human Development," *Ekistics 26* (June 1968), p. 374–394, p. 380.

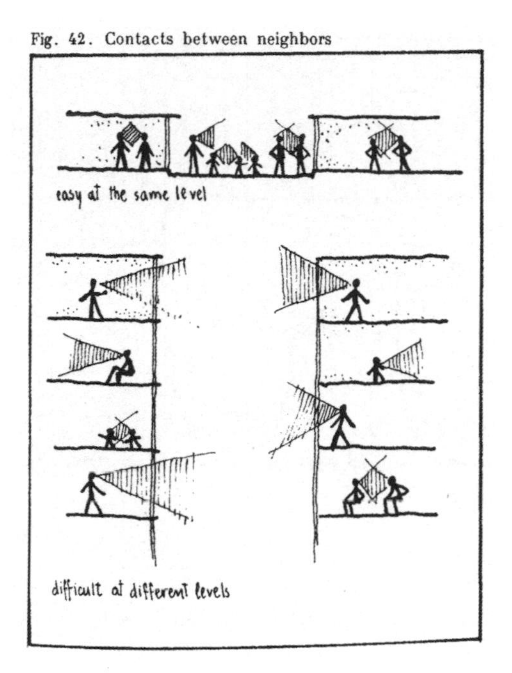

Fig. 3: C.A. Doxiadis, "A City for Human Development," *EKISTICS* (1969).

The social highlight of the annual EKISTICS month in Athens was a seven-day boat cruise on the Aegean Sea which climaxed in a grand convention among the antique ruins of the abandoned island Delos, the so called *Delos Symposium*. *Delos* was not an ordinary congress or convention. The congregation of scholars and planners was the highlight and annual meeting of the *World Society for Ekistics*. Doxiadis not only founded the society, but also the science: *Ekistics*, derived from the Greek adjective οἰκιστικός, wants the science of human settlement to play a central role in environmental design as well as in global decision- and policy making. He pursued an interdisciplinary approach towards planning, biological informed cybernetics with a focus on social and systems design of settlements, urban planning, and new technologies.[7] Being a successor to the Macy Conferences and cybernetics, the interdisciplinary society shared not only central

---

7 Ekistics has been a registered non-governmental organization since 1970 (until today), but Doxiadis claims that the concept dates back to the 1940s (See URL: http://www.ekistics.org).

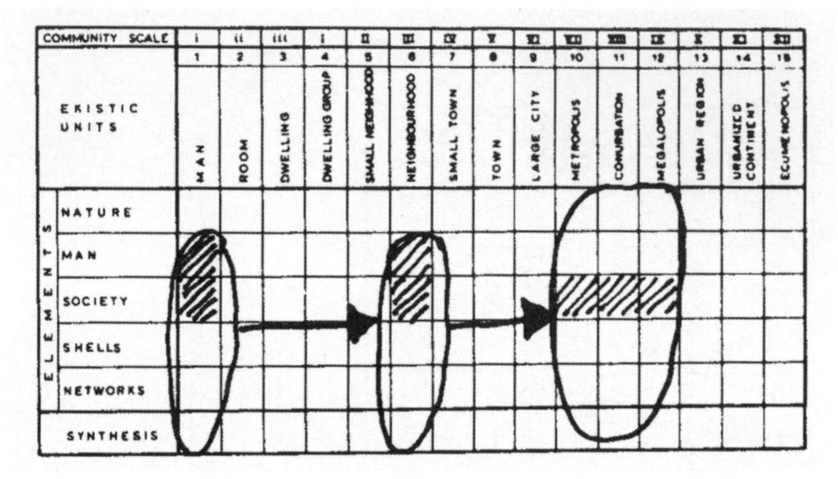

Fig. 4: EKISTIC Grid for the report of "Delos 7" (1969).

ideas on communication and regulation of systems with the Macy conferences, but also some protagonists.[8]

Held every summer between 1962 and 1974, *Delos* lasted about ten days and hosted a wide range of planners like architects, scientists, artists, and politicians, who approached problems of human settlement from various disciplinary and technological perspectives. Significant for the Delos Symposia, besides the informality and privacy guaranteed by the fact that these meetings took place offshore, was the central role of graphics, sketches, diagrams and flow charts in the discussion groups, some of which found their way into the publication of the Delos Reports. The meetings, as Mark Wigley has shown, were thoroughly documented and immediately published in the context of the monthly periodical, including lists of participants, summaries of lectures and discussions, and lecture scripts. The documentation details the international, interdisciplinary and inter-institutional frame of *Delos*, that exceeded the regular work of the *Society for* Ekistics, based in Athens, by far (Fig. 4).[9]

---

8 Above all Margaret Mead and Gregory Bateson. For an overview of participants of the Macy Conferences see Claus Pias, ed., *Cybernetics / Kybernetik. The Macy Conferences 1946–1953*, Vol. 1 (Zurich; Berlin: Diaphanes, 2003).

9 See Mark Wigley, "Network Fever," *Grey Room* 04 (Summer 2001), p. 82–122. Wigley stresses the fact that Delos was a media event. See p. 89.

Every section of the Ekistics Magazine begins with a diagram that locates the content within the cosmos of Ekistics. The diagram allows the reader to scan the range and dimension of the following text in conformity with the order of the five Ekistics elements (nature, man, society, shells and networks) and the settlement scale that is attended to in the paper, e.g. a neighborhood unit contains 1500 people, a megalopolis amounts to 100 million and the *ecumenopolis* specifies the entire world population of then 3 billions.

The Ekistics Unit for the report of Delos 7 shows how central the concept of neighborhood was for the lectures and workshops that were held in the summer of 1969, neighborhood being the key unit between the individual and the metropolis, conurbation or megalopolis. The report calls for urgent action in the face of several crises, among them the threat of nuclear war and the "world-wide crisis of urbanization", meaning that the majority of the world population "is housed in either rural hovels or in the super slums of great cities".[10]

One of the former participants was Marshall McLuhan. After his contribution, Delos 7 emphasized the promising role of new, earth-spanning transport and communication technologies such as television, satellites and last but not least the space program. In 1969, there is a new *global* aspect to neighborhood, as the human environment is now the entire earth.

One of the discussions at Delos 7 dealt with "Some techniques of identifying the problems of social groups".[11] The participants discussed the possibility of studying social systems by game theory, believing that recently developed non zero sum games would serve the purpose of studying living systems better than zero sum games. Zero-sum-games as developed by John von Neumann and Oskar Morgenstern in 1944 are games of winners and losers, supposed to reward behavior of deceit and misleading as being implemented in Cold War strategies. Non-zero-sum games are drawn from biological systems ("the heart does not win over the liver") and can promulgate cooperative behavior if the collective advantage in cooperation takes the upper hand within a series of games: "By changing the pay-off, you can change the ratio of this collective advantage. It is clear this is a real dilemma that confronts nations as well as neighbors."[12]

—

10 "Delos Report," *Ekistics* 28 (October 1969): p. 215.

11 "Some Techniques of Identifying the Problems of Social Groups. Discussion at Delos", *Ekistics* 28 (October 1969), p. 253–256.

12 "Delos Report", p. 255.

The problem is that cooperative behavior comes at a high risk for the individual or the group if the paradigm is non-cooperative. So in order to change the behavior of the players, you first have to change the paradigm. This is where Richard Buckminster Fuller's *World Game* comes into the picture.

3. WORLD GAME

When Buckminster Fuller, who participated in almost every single one of the annual cruises between 1962 and 1974, introduces his *World Game* at Delos 7, anthropologist Margaret Mead, social planners and system theorists Conrad Waddington and Karl Deutsch were on board, as well as influential entrepreneurs and politicians.[13]

In his talk, Fuller explains that the student revolt, "the uproar of the youth" as he calls it, is caused by a positive new awareness of the *global* neighborhood, "the result of a new and vivid awareness of all other humans around our space vehicle Earth, a vivid awareness of neighbors, never before experienced by humanity."[14] The fact that the student movement mainly shows negative results was simply due to the lack of positive alternatives to the present state of affairs, and *World Game* would provide these alternatives.

Fuller names a number of global challenges, among them the thread of nuclear war, environmental pollution, or the gap between rich and poor. But at the same time, he emphasizes that *World Game* could provide the technological answers and solutions for these problems. By modeling the possible outcomes of mainly resource-oriented decisions, "a series of participating simulations on how to make the whole world work better", *World Game* would produce the right solutions to all global-sized problems.

Like all of Buckminster Fuller's designs, *World Game* is difficult to describe as an object. Rather it is a process of constant redesign – game design in this case –, inspired by statistical graphics and computer-based and non-computer-based war and strategy games of his time. *World Game* promised not only to understand global dimensions of ecological, social and political factors, but to empower different global agencies.

---

13 Fuller met e.g. Walter Rostow, a professor for economics in Texas and freshly put out of office as Assistant for National Security under Lyndon B. Johnson and in this function partly responsible for Johnson's Vietnam War strategy. When Johnson did not run for a second presidency due to the public critique on the war, Rostow went back to economic science. See the picture of Fuller and Rostow in *Ekistics* 28 (October 1969), p. 255.

14 Richard Buckminster Fuller, "World Game," *Ekistics* 28 (October 1969), p. 286–292.

The tools were already at hand, but would not come cheap, as Fuller informs the *Delos* participants: "We have organized at Southern Illinois University, and we hope it will soon be in operation, a $16 million dollar computer implemented program for playing just such a mutual success seeking game in a dramatically visible way. It will be so photogenic that it will become popularly and repeatedly broadcast on the world's TV circuits. Thus society may come to realize not only what *is* happening but also what could happen in an omni-favorable way."[15]

According to Fuller and his *Delos* companions, computer based simulations would work where traditional politics were failing: "I think that possibly one of our greatest problems is the educational problem of getting man around the Earth, to realize in time what his problems are and what the most effective priorities may be for saving them all, as discovered by the computer and not as dictated by anyone's opinion or by the passionately evoked opinions of any political bias."[16]

And since the self-made man Fuller did neither believe in politics nor in waiting, he had already set up *World Game* as an open student research project in New York City, before he set off to *Delos 7*. The reconstruction of what exactly happened under the brand of *World Game* is difficult, since the game had no rules besides the one that says that either everybody wins or everybody loses. The parameters and statistical data to simulate global resource distribution and consumption were only rudimentarily available, and the project had no computer hardware to run any simulations. From the documentations one can tell that *World Game* started like most of Fuller's classes with several hours of his universal "energetic geometry", that he had to leave shortly after, and that Edwin Schlossberg, a doctoral candidate in literature and physics of Columbia University, took over. The participants were between 19 and 46 years old, they came from Chicago, San Francisco, and New York, they studies art, physics, architecture or biology. First they had to gather data from the archives of the United Nations, the *New York Times*, the Embassies or U.S. government organizations. They read books and reports, they watched film documentaries, they drew flowcharts and diagrams to follow resources like energy, food and technologies around the globe. The search for general trends and survival paradigms for a single individual on the planet reminds one of

---

15 Buckminster Fuller, "World Game," p. 287.

16 Buckminster Fuller, "World Game," p. 291.

the early think tanks of the RAND Corporation, these "institutionalized negotiations between expertise and dilettantism."[17]

The participants worked out visualizations and necessary measurements and steps for the future course of action on planet earth. The rising computer industry also showed interest. The IBM magazine *Think* covers Fuller's and Schlossberg's summer school under the headline *Today Greenwich Village, Tomorrow the World* in its November 1969 issue, and enthusiastically reports some extraordinary results, e.g. the development of an energy network based on global neighborhood and time differences: while Canada sleeps, Russia could use their excessive energy and vice versa.[18] Everybody wins.

The diary entries of playmaker Edwin Schlossberg are carried by a similar enthusiasm. They speak of two major events that took place that summer, *World Game* and *Moon Landing*:

> "July 16 – Man leaves for the moon. As we leave the earth to be able to see it in its entirety, a small group of people work to document this ability and to employ the documentation in order to quickly distribute the resources and the knowledge to all of humanity. July 17 – MOON DAY – Everybody comes in exhilarated but exhausted. [...] As man is now able to leave the earth and stand on another body in space he can see the earth for the first time as a spaceship. The students express confidence in the acceleration towards physical success for all of humanity and realize that they have been working to assure this. We are explorers. Mr. Fuller was called a lunatic by some earlier in his life. Now lunatic is good."[19]

The simultaneity of the two missions was no coincidence. Wherever Fuller turns up in 1969, he lobbies for *World Game*. In June he speaks at a joint meeting of the *American Astronautical Society* and the *Operation Research Society* in Denver to astronauts, military and scientists. In his 30-page long paper, Fuller presents himself as a former Navy man (in World War I) and refers extensively to shipbuilding and uses ship metaphors to pro-

---

17 See Claus Pias and Sebastian Vehlken, "Von der Klein-Hypothese zur Beratung der Gesellschaft," *Think Tanks. Die Beratung der Gesellschaft*, ed. Thomas Brandstetter, Claus Pias and Sebastian Vehlken (Zurich; Berlin: Diaphanes, 2010), p. 7–16, p. 13.

18 IBM, *Think* (November–December 1969). This magazine was published exclusively for the employees of IBM (Stanford Special Collections, *Buckminster Fuller Papers*, Series 2, Box 190).

19 Edwin Schlossberg, *World Game Diary* (Stanford Special Collections, Buckminster Fuller Papers, Series 18, Box 24).

mote the idea of *World Game* and synergetic and sustainable thinking among national leaders.[20]

Fuller's plea for regenerative energy among high-ranking military might come as a surprise, but from his perspective, the important aspect of a new technology is not the scientific or economic context of its development but its application *within* regenerative environmental strategies and technology. Government and military interests are just as much part of the background of *World Game* as the student and peace movement of the late 1960s or the beginning excitement in market simulations of industry.

The Southern Illinois University (SIU), where Fuller (who never actually graduated from a university) was a faculty member in those years, advertised *World Game* as part of a new design and planning science, which recognizes the central role of data and information processing: "World Gaming is an engineered attempt to plug in our sensory awareness mechanisms to the switchboard of *universe*, to get in sync witch the metabolism of this spaceship's environment. [...] Planning, any kind of planning depends upon information input (and on being aware of the proper question to ask of the data)."[21]

For the environmental synchronization to happen, one needs a game environment where decisions can be played through without risk. The players are supposed to train their thinking for the possibilities of alternative actions, to broaden their action horizon for future decisions. The five-year-planning study of SIU also refers to the educational concept of the Toronto School of Marshall McLuhan and others: It focusses on visual media and the idea of a learning machine, centering around an interface or display which immerses the audience in data processing:

> "The possibility of arranging the entire human environment as a work of art, as a teaching machine designed to maximize perception and to make everyday learning a process of discovery (McLuhan: The Medium is the Message, pg. 68)[sic!]. The beneficiary of the environment (the planner, decision-maker, administrator, politician) must be confronted by a total environment for planning – an environment, which will expand his awareness

---

20 Richard Buckminster Fuller, *World Game (Preprint)*. Joint National Meeting American Astronautical Society and Operations Research Society, June 17–20, 1969, Denver, Colorado (Stanford Special Collections, Buckminster Fuller Papers, Series 18, Box 25). For the ship metaphor and Fuller's navy connections see also Peter Anker, "Buckminster Fuller as Captain of Spaceship Earth," *Minerva* 45 (2007), p. 417–434.

21 Tom Turner, *Report of World Game* (Stanford Special Collections, Buckminster Fuller Papers, Series 18, Box 24–25). Shortly after Turner's report, Turner and Fuller had a massive argument and Fuller left SIU in 1971 (see Hsiao-Yun Chu, "The Archive of Buckminster Fuller," *New Views on Buckminster Fuller*, ed. Roberto Tujillo (Stanford, CA: Stanford University Press, 2009), p. 20.

> of possible decision alternatives to spaceship resource utilization in much the same way as our conception of *universe* was expanded by seeing for the first time the sapphire earth against the black void of space – *terra firma* passed into history. [...] World Gaming will, like fashion, engineer an environment, which seeks to continually enhance sensory inputs to long-range planning strategies of decision-makers by as graphically and hypervisually, as skill and art will allow, presenting resource information in a total world, audience enveloping display."[22]

In its design, *World Game* was a prototype for the North-American media ecological paradigm which went viral within theoretical discourse and was implemented in concrete media and information technologies; a global neighborhood design program based on the interconnections of media technologies, education utopia and gouvernmentality.[23] *World Game* operates in the transition zone between science and art.

4. WORLD SIMULATION

The prehistory of *World Game* began, as did the notion of cities as living organisms, in the early 1960s, when Fuller began to work on the *World Resource Inventory* in collaboration with artist and sociologist John McHale. The six volumes – jam-packed with tabloids and number columns – that they published between 1963 and 1967 are intended to be a global inventory of resources and trends to enable an ecological redesign of the planet.[24]

When Fuller announced the *World Design Science Decade* in 1961, he addressed future architects, urban planners and environmental designers. In his vision, the educational institutions at universities and liberal art colleges would reorient towards world planning and world design, and the *World Resource Inventory* would serve as their database. The project might have been the first civil computer based big data project ever, "a coordinating agency, which acquires and disseminates information [...] and provides guide

---

22 Turner, *Report of World Game.*

23 The term *media ecology* was coined by Neil Postman, using a biological metaphor: the Petri dish which reveals "that a medium was defined as a substance in which a culture grows. If you replace the word 'substance' with the word 'technology', the definition would stand as a fundamental principle of media ecology" (Lance Strate, *Echoes and Reflections: On Media Ecology as a Field of Study* (Cresskill, New Jersey: Hampton Press 2006), p. 15).

24 See Joachim Krausse, ed., *R. Buckminster Fuller. Your Private Sky. Diskurs*, Vol. 2 (Zurich: Müller, 2001), p. 263.

analysis of the basic world trends data to the various school projects around the world" – a predecessor of the *Limits to Growth* report to the *Club of Rome*, published in 1972.[25]

In order to turn the database into a decision-making machine, in order to turn the *World Resource Inventory* into a *World Resource Simulation Center*, Fuller needed money for a super computer. He saw his chance to develop his *total environment for planning* when he was asked to design the U.S. pavilion for the World Exhibition *Expo 1967* in Montreal. Fuller's first draft reminds one of a gigantic football stadium, built around a Fuller world map as a dynamic, electronic game field for displaying resource and population distribution, a display for possible futures of the Cold War.[26] But the U.S. information agency rejected the concept, and in the end Fuller was only allowed to design the building. However, this turned out to be his greatest success as an architect. What should have been a gigantic simulation environment to teach an international audience the interdependence of political and economic decisions on a global scale became a demonstration of American superpower and Fuller's architectural masterpiece, the *Expo Dome*.[27] After the world exhibition Fuller used his publicity to continue lobbying for a super computer at Southern Illinois University, a computer that "should have replaced all human capacity."[28] The costs for the center were estimated around 16 billion dollars, according to a study of the university in 1967.[29]

In his view, the center should focus on the development of new technologies, electronic computers, game theory and simulations, its aim being the optimization of a worldwide resource management. The study states that no comparable facility existed. Although all single tasks and applications already existed at different places in the U.S., the main idea was to concentrate the simulation knowledge from military, government

---

25 See John McHale, *World Trends Exhibit*, introduction to an exhibition catalogue that was shown 1967 at the 9th World congress oft he U.I.A (Union of International Architects) in Prague, at the conference "Man and his Environment" in London, at the conference of the Design Science Decade Day of Expo 1967 in Montreal, at a conference on Alternative Futures in Washington, D.C., in 1968, and during a seminar on International Affairs at Princeton University. For the future scenarios of the Club of Rome, see Donatella Meadows et al., eds., *The Limits to Growth. A Report for the Club of Rome's Project on the Predicament of Mankind*, (New York: Universe Publishing, 1972). If there actually was a direct connection between Jay Forrester and Buckminster Fuller is unclear, but both were except Delos participants.

26 See Christina Vagt, "Fiktion und Simulation. Buckminster Fullers *World Game*," *Mediengeschichte nach Friedrich Kittler*, ed. Friedrich Balke, Bernhard Siegert and Joseph Vogl (Paderborn: Wilhelm Fink Verlag, 2013), p. 117–134.

27 See Krausse, *Buckminster Fuller. Your Private Sky*, p. 422, p. 469.

28 Anker, "Buckminster Fuller as Captain of Spaceship Earth," p. 428–431.

29 Geometrics Inc. (Ed.), *Preliminary Plan Study for the Centennial World Resources Center for Southern Illinois University*, September 1967 (Stanford Special Collections, Buckminster Fuller Papers, Series 18, Box 24–25.)

institutions, scientific and industrial research institutes, in the form of computers and know-how, at Southern Illinois University.

The challenge of the project was not only to acquire the necessary hard- and software solutions, but also the headhunting of the estimated 60 employees, since "these people tend to concentrate in the East and West coast sections of intensive research activity" and not in Southern Illinois. A new building and a time-sharing computer system with large memory capabilities had to be financed, software had to be developed. Where technological performance did not yet meet the expectations, e.g. in the case of big data processing computing machines or the performance of large reactive displays, Fuller's team searched for interim solutions. "In these circumstances, the exactness of the data is not critical. [...] The main argument for their use at the present time is in terms of learning situations. This is to say that the process of the simulations or game is the important aspect, not the end product".[30]

The center should have be about the development of methods, research, education- and publicity work, a gigantic interface for education, research, industry and media technology.

The 5-years-study does not predict however, whether the center, even if it had been fully financed, would have been able to run global resource- and trend simulations by 1974. In the end, the center was not realized. Thus, SIU never acquired a super computer that was continuously fed with worldwide data, and *World Game* never evolved into the global information network whose computers displayed the outcome of data retrieval, aerial image and satellite scans of earth's flora and fauna to everyone and everywhere. Nonetheless, Fuller never got tired of talking about how *World Game* was already running, that it had already been implemented for decades, not in the form of computer hardware, but in the form of the *metaphysical software* of the human mind.

World Game was, like probably most simulations of the 1960s, not a medium of prediction, but a medium to educate and control society, a medium to change the future. Or, in other words, it was "a total environment for planning" that conditioned the behavior of its players according to Fuller's motto: "Reshape environment; don't try to reshape man."[31]

Although Fuller stresses the fact that *World Game* should be seen as an alternative to actual politics, the archives give notice of the rather conventional political strategies it already served: In 1970, the 91st congress of the U.S. was debating a bill introduced by

—

30 Geometrics Inc., Preliminary Plan Study, p. 4.

31 Krausse, *Your Private Sky*, p. 218.

Congressman Melvin Price. This proposed act would have authorized NASA to fund Fuller's and McHale's *World Resources Simulation* Center at SIU, allowing governmental and non-governmental institutions to gather information for decision making and planning. Price argued that basic data for the new productive information tool had already been prepared by Buckminster Fuller's *World Game* Program, but more important, *World Game* had already proven that simulation games were a new kind of governance: "It is self evident that young people in every State in the Union are groping for new directions in this troubled world, and I am happy to report that on their own volition students and teachers are already engaged in the World Game and resource simulations."[32]

5. SKYRISE FOR HARLEM

While Fuller was involved with the design of global simulation and planning tools, he was at the same time working on a much smaller neighborhood design scale. As Cheryl J. Fish has shown, Fuller drafted a redesign for New York City's main ghetto Harlem in collaboration with African American poet and activist June Jordan and architect Shoji Sadao between 1964/65.[33] Jordan addressed Fuller shortly after the Harlem riot in 1964, the first in a series of hundreds of riots in the black neighborhoods of NYC between 1964 and 1968, asking for an environmental redesign for the neighborhood that could provide solutions for a situation that had been called "the most direct challenge ever posed to the American social order, an order historically based upon racial discrimination and ethnic fragmentation among the lower classes".[34] Instead of clearing the slum, Fuller's and Sadao's redesign of a community of 250,000 people in northern Manhattan was based on the idea of moving Harlem's tenants *up* – by building the new skyscraping high-rise dwellings above the old and creating large areas of recreation and parking.[35]

Although the redesign was never realized either, it's case shows the epistemic shift from urban planning to environmental design between early 1960s and 1970s, with the neighborhood as the central social unit, and how it stems from the cybernetic paradigm of controlling a population by regulating its environmental conditions.

—

32 91. US Congress, 2nd session, H.R. 17467, House of Representatives, May 6, 1970 (Stanford Special Collections, Buckminster Fuller Papers, Series 18, Box 24–25).

33 Cheryl J. Fish, "Place, Emotion, and Environmental Justice in Harlem: June Jordan and Buckminster Fuller's 1965 *Architextual* Collaboration," *Discourse Vol. 29, No. 2&3* (Spring & Fall 2007): p. 330–345.

34 Fish, "Place, Emotion, and Environmental Justice in Harlem," p. 339.

35 Fish, "Place, Emotion, and Environmental Justice in Harlem," p. 340.

HENRIETTE BIER

DIGITALLY-DRIVEN DESIGN AND ARCHITECTURE

ABSTRACT

Distributed, networked, electronically tagged, interactive devices are increasingly incorporated into the physically built environment, progressively blurring the boundary between physical and virtual space. This changing relationship between the physical and virtual domains implies not only a change in the operation and use of physically built space but also in its physical configuration, and therefore, its design. Architecture incorporating aspects of intelligence[1] employs information and knowledge contained within the network connecting electronic devices.[2] Thus, the relevant question is not whether intelligent, sentient environments may be built, but how these environments may become instruments for distributed problem solving and how (artificial) intelligence[3] may be embedded into architecture in order to serve everyday life.

In this context, digitally-driven architecture is defined as an architecture that is not only designed and fabricated by digital means but which, actually, incorporates digital sensing-actuating mechanisms[4] that enable buildings to interact with their environment and users in real-time. This paper discusses[5] digitally-driven design and architecture[6] that incorporates on some level bottom-up mechanisms enabling the emergence of global effects from local interactions. While digitally-driven architectural design may imply the emergence of spatial and programmatic formations from contextual (environmental, programmatic, etc.) interactions, digitally-driven architecture employs

—

1 Eli Zelkha, Brian Epstein, Simon Birrell and Clark Dodsworth "From Devices to Ambient Intelligence," *Digital Living Room Conference*, http://epstein.org/brian/ambient_intellegence.html (accessed September 2014).

2 Emile Aarts, Rick Harwig and Martin Schuurmans, "Ambient Intelligence," *The Invisible Future: The Seamless Integration of Technology Into Everyday Life* (New York: McGraw-Hill, 2001).

3 Stuart Russell and Peter Norvig, *Artificial Intelligence - A Modern Approach* (New Jersey: Upper Saddle River, 2003), p. 1–40.

4 Henriette Bier and Terry Knight "Digitally-driven Architecture," *Footprint* 6th issue, ed. Henriette Bier and Terry Knight (Delft: TU Delft, Stichting Footprint, 2010).

5 Henriette Bier, "Building Relations," *Architectural Annual 06* (2005), ed. Bekkering (Rotterdam: 010-Publishers), p. 64–67.

6 Bier and Knight, "Digitally-driven Architecture".

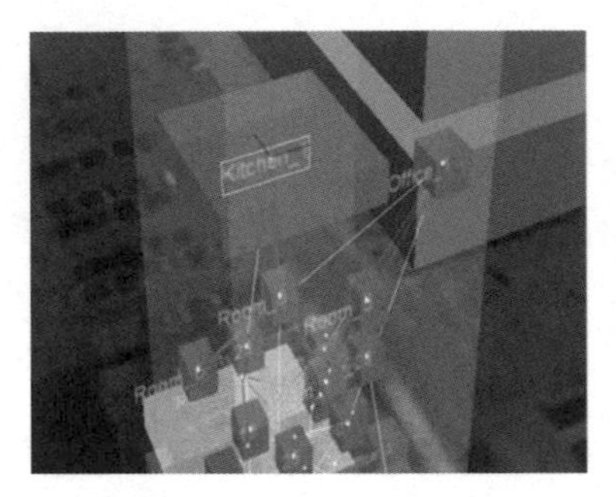

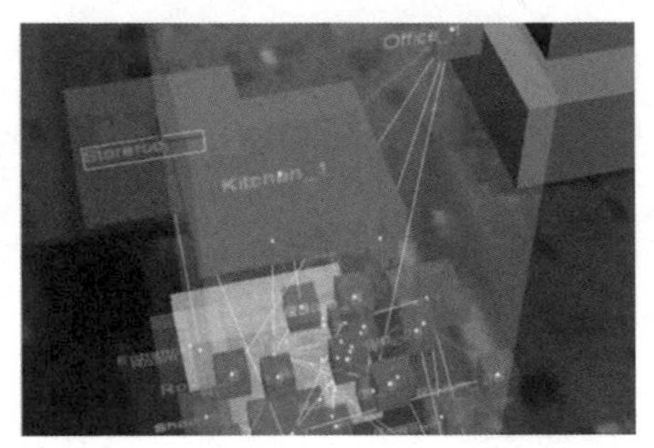

Fig. 1: Building Relations – Architectural design implemented in multi-agent environments developed in Virtools (2005).

real-time interaction in the actuation of architectural embodiments, which become dynamic, acting and re-acting in response to environmental and user-specific needs.

1. DIGITALLY-DRIVEN DESIGN

Digitally-driven design processes employ parametric design in production chains that are establishing a direct link between design conceptualization, process, and result. Such digitally-driven design processes have been the focus of current architectural research and practice inter al. due to the phenomenon of *emergence* explored within self-organization, generative grammars, and evolutionary techniques. Self-organizing swarms,[7] for instance, are employed in generative design processes, which deal with ample amounts of data featuring sometime conflicting attributes and characteristics. Those attributes and characteristics are incorporated in behaviors according to which design components such as programmatic units swarm towards targeted spatial configurations (Fig. 1). In this context, architectural and urban design become procedural instead of object-oriented, whereas architectural and urban form emerge in a process of interaction between all parts of the system. Thus, the architect becomes the designer of a process and only indirectly of a result.

Such swarms operate as multi-agent systems and consist of simple agents that interact locally with one another and their environment based on simple rules leading to the

---

7 Henriette Bier, "Building Relations," *Architectural Annual* 06 ed. Bekkering (Rotterdam: 010-Publishers, 2005), p. 64–67.

emergence of complex, global configurations. Their use in design is of relevance because of their ability to embody both natural (human) and artificial (design-related) aspects. In the context of urban and architectural design, swarms of agents consist of both natural (human) and artificial (software) agents that interact with each other and the environment. Such systems are set up as parametric models incorporating characteristics and behaviors representing the natural and artificial systems themselves, whereas simulations of behaviors show operation of such systems in time.

Simulations are of interest, in this context, not so much for their ability to represent (and confirm) assumptions or even improve (optimize) design solutions but for their generative potential based on emergence. This implies that the design emerges from a process of self-organization, in which the dynamics of all parts of the system (agents and environment) generate the result. Such generative processes implemented in simulations (Fig. 2) are extensively discussed inter al. by De Landa[8] in relation to the Deleuzian understanding that matter itself has the capacity to generate form[9] through immanent, material, morphogenetic processes. Thus design as production of representations of artefacts (by means of drawing, modelling and simulation) implies systemic interaction between (human and non-human) system components[10] while authorship increasingly becomes hybrid, collective, and diffuse.

Interactions between human and artificial agents may follow principles as described in the Actor-Network Theory (ANT) implying that material-semiotic networks are acting as a whole[11] whereas the clusters of actors involved in creating meaning are both material and semiotic. ANT, therefore, does not differentiate between human and nonhuman actors since differences between them are generated in the network of relations, implying agency of both humans and non-humans, whereas agency is neither located in human subjects nor in non-human objects, but in the heterogeneous associations between the two. This understanding is extensively discussed in De Landa's *new*-materialist cultural theory[12] that rejects the dualism between nature and culture, matter and

---

8 Manuel De Landa, *Philosophy and Simulation: The Emergence of Synthetic Reason* (New York: Bloomsbury Academic, 2011), p. 111–145.

9 Manuel De Landa, *Materialism, Experience and Philosophy* (European Graduate School, 2008), http://www.youtube.com/view—play—list?p=38A848FA0C7479C3 (accessed March 2013).

10 Kas Oosterhuis and Henriette Bier, *Robotics in Architecture* (Heijningen: Jap Sam Books, 2013).

11 Bruno Latour, *Reassembling the Social: An Introduction to Actor-Network-Theory* (Oxford: Oxford University Press, 2005), p. 63–86.

12 Rick Dolphijn and Iris van der Tuin, *New Materialism: Interviews & Cartographies* (Michigan: MPublishing, University of Michigan Library, 2012), p. 38–47.

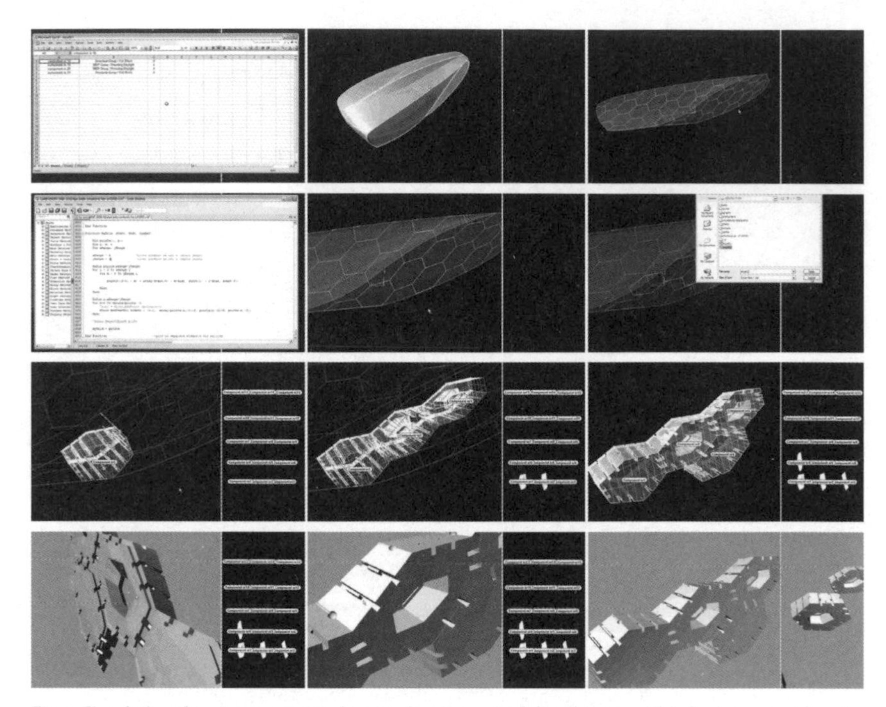

Fig. 2: Simulation showing parametric layout of componential distribution within Protospace 4.0 (2009).

mind, natural and artificial, in which reality is considered to reveal itself in material, self-organized processes.

Interactive design tools developed at *Hyperbody* in the last decade employ swarm intelligence and consist of software agents implemented sometimes as functional units interacting locally with one another and with their environment[13] as follows: Programmatic units pertaining to a building or neighborhood are defined as flocking agents striving to achieve a preferred spatial layout. In this context, spatial relations between programmatic units can be described as rules, according to which all units organize themselves: While the designer might find it difficult to have an overview of all functions and their attributed volume and preferential location, functional units can easily *swarm* towards locally optimal configurations.

—

13 Henriette Bier and Yeekee Ku, "Generative and Participatory Parametric Frameworks for Multi-player Design Games," *Footprint* 13th issue, ed. Maros Krivy and Tahl Kaminer (Delft: TU Delft, Stichting Footprint, 2013), p. 153–162.

Such optimal configurations emerge from swarm-like approaches employing digital tools in order to design *neighborhoods* in the spatial sense either at building or city scale. Swarm-like approaches employed at *Hyperbody* change the spatial configuration of the resulting building or city neighborhood at the level where functional or programmatic distribution is achieved bottom-up; however, the formal configuration still requires the top-down input of the designer. The obvious advantage of such an approach lies in the ability to generate multiple design versions exhibiting a range of local optima from which the designer can choose a preferred one.

As programmatic layout deals with the placement of functions in 3D-space, software prototypes developed at *Hyperbody* (Fig. 1) rely on a simple strategy: Spatial units establish relationships with other spatial units by determining their distance to each other and automatically adjusting their width, length, and height in order to prevent potential misplacements, overlaps, and collisions. Programmatic units, therefore, adjust themselves to their surroundings, they are linked to other units creating spatial relations defined and simulated by taking a program of requirements (the number of specific functions, their volume and occupancy numbers, etc.) and translating them into organized spatial layouts. Such layouts are achieved by defining minimum-maximum distances between objects such as units and surroundings based on rules of attraction and repulsion.

These self-organization mechanisms are complemented by interactivity, as the layout process does not take place outside of the influence of experts and users, who can directly select and move objects, adjust parameters while the simulation re-adjusts to the new input values. In this way, interacting artificial and natural (experts and users) agents search for preferred programmatic configurations, whereas the users' choice is limited to a range of high and low density, high- and low-rise typology, and diverse-hybrid or mono-homogeneous programmatic functionality predefined by experts.

Such generative, interactive design tools continuously receive and send data from and to a database, which contains all information regarding programmatic units. The units thus are defined by type, function, scale and position, 24/7 use, etc. Other design related sub-tools running in parallel might use these values or combinations of values in order to allow experts to investigate structural, formal, environmental implications. These tools are, therefore, used interactively and in combination with other software, in order to achieve locally optimized designs. Despite their diagrammatic character, these applications demonstrate an obvious capability to support the functional layout of large and complex architectural and urban environments based on emergent swarm principles.

In such a networked system users are connected with other users, multimedia databases and applications enabling the reading and editing of data, sensing-actuating, and computing in such a way that users interact physically and virtually as needed in a physical, digitally-augmented environment. Similarly, at architectural scale interaction between physically built spaces, indoor-outdoor environment and users establish connectivities between virtual and physical environments.

2. DIGITALLY-DRIVEN ARCHITECTURE

The ongoing fusion of the physical and the virtual reflected in the convergence of the Internet, mobile communication systems, and advanced human-computer interaction technologies generates a physical *reality-virtuality continuum*[14] containing all possible degrees of real and virtual conditions so that the distinction between physical reality and virtuality becomes blurred. The relevant question is, therefore, not if but how the virtual and the physical interface architecture, and how (artificial) intelligence may be embedded into physical environments in order to serve everyday activities. The assumption is that buildings operate within the spatial conditions of the urban territory as a feedback system negotiating (intelligently) between (artificial or natural) physical and virtual components of the networked system.

Thus, digitally-driven architecture has been implemented by Hyperbody in physically-built prototypical environments (Fig. 3) that accommodate, on the one hand, human needs addressing imperative requirements for responsiveness, adaptation, flexibility, and reconfiguration. On the other, they extend human needs by establishing interactive relations with the environment. Reconfigurable, robotic environments incorporating digital control namely sensor-actuator mechanisms that enable buildings to interact with their users and surroundings in real-time through physical or sensory change and variation allow multiple, changing functions in condensed time frames[15] and for instance address local issues of inefficient use of built space but also global issues of population migration with respect to humanitarian aid in catastrophic conditions and rapid urbanization due to population density growth.

—

14 Paul Milgram, Haruo Takemura, Akira Utsumi and Fumio Kishino, *Augmented Reality: A class of displays on the reality-virtuality continuum*, proceedings of SPIE vol. 2351, ed. Hari Das (Bellingham: SPIE, 1994), p. 282–292.
15 Bier and Knight, "Digitally-driven Architecture".

Fig. 3 InteractiveWall developed by Hyperbody in collaboration with Festo responds to people's movement (2009).

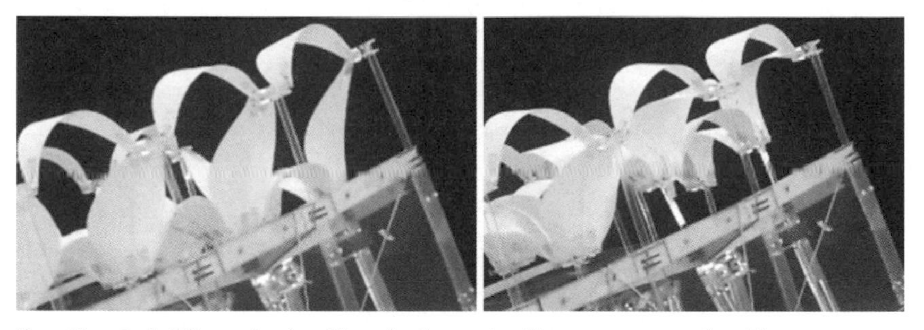

Fig. 4: Hyperbody MSc 4 project (2012) featuring interactive skin components employed for energy generation and ventilation purposes.

Application of embedded robotics in architecture implies, therefore, allowing for downtime (referring to time periods when the system, in this case the building, is non-operational) to be reduced through physical reconfiguration. This is accomplished through the advancement of collective behavior systems so that several autonomous building components operate in cooperation in order to accomplish major reconfiguration and adaptation tasks. The aim is to address societal issues such as the current inefficient use of built space (as most buildings are only used 8–16 from 24 hours per day) through spatial reconfiguration. Last but not least, the advancement of embedded, interactive or robotic, energy systems (Fig. 4) employing renewable energy sources such as solar and wind power are aiming at reducing architecture's ecological footprint,

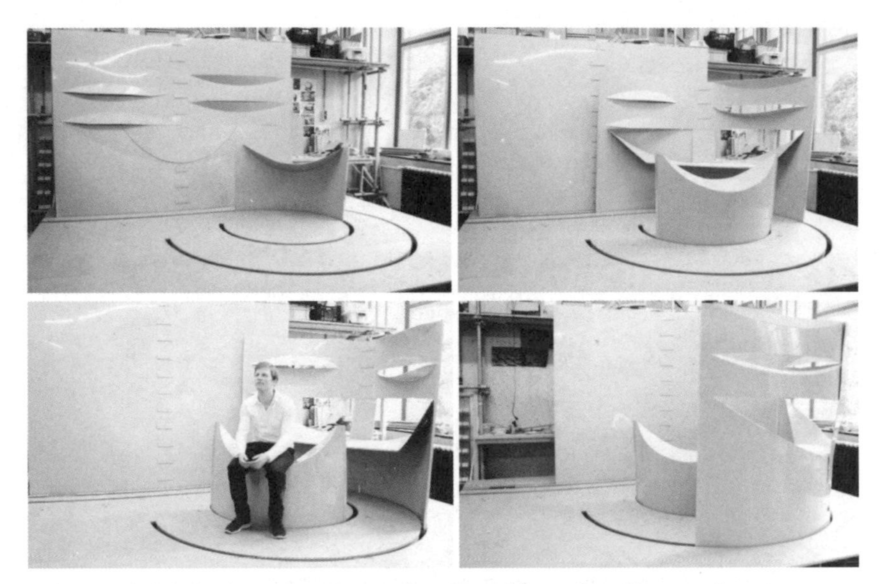

Fig. 5: Hyperbody MSc 2 project (2013) featuring a multi-modal, reconfigurable system for an apartment.

whereas distributed robotic climate control allows energy efficient human-centered and demand-driven indoor climate regulation.

In this context, digitally-driven architecture exploits swarm intelligence at the level where climate control devices are embedded in building components as shown in the MSc 4 project (2012) featuring interactive skin components employed for energy generation and ventilation purposes.

Furthermore, reconfiguration is explored in projects such as *Multimodal Apartment*: It is aimed at the development of a small apartment of 50 m^2/150 m^3 that has all the spatial qualities and functional performances of a standard 100 m^2/300 m^3 apartment. For that purpose, the design had to cater for radically different kinds of use during the course of a day, a week, or a year. The assumption was that when one is in the living room, one does not use the bedroom, and when one is cooking, one does not use the bathroom at the same time. At one moment of the day large sections of the space could cater to only one or two functions. The Pop-up Apartment (Fig. 5) combined material and geometrical properties with robotic features in order to enable spatial reconfiguration by moving components along a rail so that surfaces bend automatically and create spatial enclosures or pieces of furniture. Such a reconfigurable structure enables flexible use of built space implying that neighborhoods could cater to a wide range of changing activities.

These projects demonstrate that networked structures incorporating digitally-driven sensor-actuator devices exhibit behaviors of a *swarm* enabling flexible and dynamic change in varying time frames. These behaviors might be programmed to address a multitude of needs or goals from personal to societal, from aesthetic to functional, from emotional to environmental.

3. CONCLUSION

Digitally-driven, agent-based design processes imply that the same or similar (virtual and physical) agent systems may produce multiple (and even endless) variations of architectural configurations under similar conditions due to the emergent properties of the system. While digitally-driven architectural design implies the emergence of spatial and programmatic formations from contextual (environmental, programmatic, etc.) interactions, digitally-driven architecture employs real-time interaction in the actuation of architectural embodiments, which become dynamic, acting and re-acting in response to environmental and user-specific needs. Both employ swarm intelligence for generative and reconfiguration purposes, respectively implying that not only design emerges from local interactions between human and non-human agents but also physically-built space adapts and reconfigures locally according to human and environmental needs.

Thus, *neighborhood technologies* based on swarm intelligence enable integration of interactive principles into architectural design to production and operation processes while exploiting synergy effects between the disciplines involved (architecture, computer science, and robotics) and rejecting the historically established dualism between nature and culture, matter and mind, natural and artificial. *Neighborhood technologies* in this context offer simulation tools for the bottom-up generation of virtual *neighborhood designs* from which the physical production and operation of reconfigurable *built neighborhoods* emerge.

ACKNOWLEDGEMENTS

This paper has profited from the contribution of the *Hyperbody* team as well as the MSc students involved in the featured projects.

III. NEIGHBORHOOD SOCIETIES

DIRK HELBING

ECONOMICS 2.0

THE NATURAL STEP TOWARDS A SELF-REGULATING, PARTICIPATORY MARKET SOCIETY [1]

ABSTRACT

Despite all our great advances in science, technology and financial innovations, many societies today are struggling with a financial, economic and public spending crisis, over-regulation, and mass unemployment, as well as lack of sustainability and innovation. Can we still rely on conventional economic thinking or do we need a new approach? Is our economic system undergoing a fundamental transformation? Are our theories still doing a good job with just a few exceptions, or do they work only for *good weather* but not for *market storms*? Can we fix existing theories by adapting them a bit, or do we need a fundamentally different approach? These are the kind of questions that will be addressed in this paper. I argue that, as the complexity of socio-economic systems increases, networked decision-making and bottom up self regulation will be more and more important features. It will be explained why, besides the *homo economicus* with strictly self-regarding preferences, natural selection has also created a *homo socialis* with other-regarding preferences. While the *homo economicus* optimizes his own prospects in isolation, the decisions of the *homo socialis* are self-determined, but interconnected, a fact that may be characterized by the term *networked minds*. Notably, the *homo socialis* manages to earn higher payoffs than the *homo economicus*. I show that the *homo economicus* and the *homo socialis* imply a different kind of dynamics and distinct aggregate outcomes. Therefore, next to the traditional economics for the *homo economicus* (*economics 1.0*), a complementary theory must be developed for the *homo socialis*. This economic theory might be called *economics 2.0* or *socionomics*. The names are justified, because the Web 2.0 is currently promoting a transition to a new market organization, which benefits from social media platforms and could be characterized as a *participatory market society*. To thrive, the *homo socialis* requires suitable institutional settings such a particular kinds of reputation systems, which will be sketched in this paper. I also propose a new kind of

---

1 This article was first published in *Evolutionary and Institutional Economic Review* 10/1 (2013), p. 3–41. We thank the author and publishers for granting us the copyright to reprint it in this volume.

money, so-called *qualified money*, which may overcome some of the problems of our current financial system. In summary, I discuss the economic literature from a new perspective and argue that this offers the basis for a different theoretical framework. This opens the door for a new economic thinking and a novel research field, which focuses on the effects, implications, and institutional requirements for global-scale network interactions and highly interdependent decisions.

1. INTRODUCTION

In the past decades, our world has changed more quickly than ever. Globalization and technological revolutions have created a world with many systemic interdependencies, high connectivities, and great complexity. The nature and intensities of many 21st century problems are an immediate consequence of this.[2] However, we lack a *Global Systems Science* to understand the world we have created, and many of our institutions are conceptually outdated.

In fact, the intellectual framework of the current institutions of our socioeconomic system is around 300 years old, for example, the work of Adam Smith (1776) on the *Nature and Causes of the Wealth of Nations*. Many attribute to him the idea of an *invisible hand* that coordinates individual interests by self-organization. Specifically, it is often believed that optimally self-regarding behavior, as assumed by the standard microeconomic paradigm of the *homo economicus*, would create maximum social welfare. This idea goes back to the so-called *Fable of Bees*[3] and is the basis of neoclassical economics.[4] Even today, many policies are founded on it, including those that demand homogeneous and free global markets.[5]

In fact, it can be shown that the *invisible hand* works well, if all conditions assumed by the First Theory of Welfare Economics are fulfilled.[6] However, under many realistic conditions at least one of the assumptions is not satisfied so that the welfare-optimal

---

2 Dirk Helbing, "Globally Networked Risks and How to Respond," *Nature* 497.7447 (2013), p. 51–59.

3 Bernard Mandeville, *The Grumbling Hive: Or, Knaves Turn'd Honest* (1715), appeared as book in 1714.

4 Note, however, that the bees belonging to a hive have basically identical genes, such that their interests are expected to be aligned from the perspective of evolutionary competition.

5 This includes, for example, the World Bank and the International Monetary Fund.

6 Kenneth Arrow and Gérard Debreu, "Existence of an Equilibrium for a Competitive Economy," *Econometrica* 22.3 (1954), p. 265–290; Gérard Debreu, *Theory of Value: An Axiomatic Analysis of Economic Equilibrium* (New Haven: Yale University Press, 1959).

outcome is not guaranteed. This is, for example, the case if individuals interact in networks rather than through one homogeneous market, or if externalities or transaction costs matter.[7] Such conditions can lead to *market failures*. The same applies to information asymmetries or cases with powerful monopolists or oligopolists.[8]

It is therefore understandable that, as Thomas Hobbes pointed out, a selfish species would require regulation. In order to create socio-economic order, he proposed that a *Leviathan* be created – a powerful state to tame the *selfish beast*.[9, 10] Even today, this constitutes the intellectual framework of the need for top-down regulation, as it also shows up in Keynesianism. So, is the idea of a decentrally self-organizing economy flawed? On the contrary, we have just not learned, of how self-regulation, i.e. bottom-up regulation, can work.[11]

The idea of self-organization is still extremely appealing and timely.[12] Ants, for example, work in an entirely self-organized and largely decentralized way without a hierarchical system of command. The same applies to social animals like termites, flocks of birds, or schools of fish. The ecological and immune systems also function in a decentralized and highly efficient way, due to the evolutionary principles of mutation and selection. Social norms are success principles building on decentralized mechanisms as well. I will argue that we can learn from principles like these to make the *invisible hand* work, such that individual and social benefits can be simultaneously reached.

It is a great scientific challenge to find principles of self-regulation that work under less optimistic conditions than those assumed by the Theorems of Welfare Economics. In particular, they should work in case of social dilemma situations, where cooperative behavior would be beneficial for everyone, but where exploiting others can create even higher individual payoffs. I would like to point out that social dilemma situations are

—

7 Ronald Coase, "The Problem of Social Cost," *The Journal of Law & Economics* 3.1: 1 (1960); Carlos Roca, Moez Draief and Dirk Helbing, "Coordination and Competitive Innovation Spreading in Social Networks," *Social Self-Organization*, ed. Dirk Helbing (New York: Springer, 2012).

8 George A. Akerlof, "The Market for 'Lemons': Quality Uncertainty and the Market Mechanism," *The Quarterly Journal of Economics* 84 (1970), p. 488–500; Andreu Mas-Colell, Michael D. Whinston and Jerry Green, Microeconomic theory (New York; Oxford: Oxford University Press 1995).

9 Thomas Hobbes, *Leviathan, or the Matter, Forme, and Power of a Commonwealth, Ecclesiasticall and Civil*, ed. Ian Shapiro (New Haven: Yale University Press, 2010).

10 The famous quote *homo hominis lupus* compares humans with wolves.

11 Here, the term *self-organization* is used for the emergence of collective behavior or properties, which may be desirable or not. *Self-regulation* is used for a set of rules that supports the adaptive self-organization of a favorable outcome such as the convergence to a system-optimal state.

12 Bert Hölldobler and Edward O. Wilson, *The Super Organism: The Beauty, Elegance, and Strangeness of Insect Societies* (New York: W. W. Norton & Company, 2008).

expected to be common in socio-economic systems, since opportunities to reach higher benefits through overpriced low-quality products or services or by misleading information exist in many economic exchange situations. For example, the value of a product might be lower, or the risk of a financial derivative might be higher than claimed. Under such conditions, cooperation can easily erode, leading to *tragedies of the commons*. Well-known examples are environmental pollution, overfishing, global warming, free-riding, and tax evasion. The breakdown of trust and cooperation may also be seen as the main cause of the financial meltdown of 2008, i.e. it could also be interpreted as the *tragedy of the commons*.[13]

Therefore, to be successful, societies must be able to deal with *social dilemma situations*. Unfortunately, the self-regarding behavior of a *homo economicus* (as reflected by a *best response rule* in our later discussion), is destined to lead to *tragedies of the commons*. The current solution to this is to introduce taxes and regulations that change the nature of the interactions and eliminate the occurrence of social dilemma situations.[14] But is this approach effective, and is it the best solution? The de facto failure of the international carbon tax raises doubts that it does. So, is a totally different approach required?

It is often assumed that other-regarding behavior tends to be costly and to create a personal disadvantage. If so, due to the merciless forces of natural selection, nothing other than a self-regarding *homo economicus* should exist. In Sect. 3, I will present computer simulations that test this assumption. Surprisingly, biological evolution can also be shown to produce a *homo socialis* with other-regarding preferences, which may overcome tragedies *of the commons*.[15]

---

13 Dirk Helbing, "Globally Networked Risks and How to Respond".

14 Dirk Helbing and Sergi Lozano, "Phase Transitions to Cooperation in the Prisoner's Dilemma," *Physical Review* E 81.5:057102 (2010).

15 Note that Adam Smith himself believed more in a competitive, but other-regarding human being, than in a maximising self-regarding *homo economicus*. In his book *The Theory of Moral Sentiments* (London: A. Millar, 1759) he writes: "How ever selfish man may be supposed, there are evidently some principles in his nature, which interest him in the fortune of others, and render their happiness necessary to him, though he derives nothing from it. Of this kind is pity or compassion, the emotion which we feel for the misery of others, when we either see it, or are made to conceive it in a very lively manner. That we often derive sorrow from the sorrow of others, is a matter of fact too obvious to require any instances to prove it; for this sentiment, like all the other original passions of human nature, is by no means confined to the virtuous and humane, though they perhaps may feel it with the most exquisite sensibility. The greatest ruffian, the most hardened violator of the laws of society, is not altogether without." Considering this, one might conclude that Adam Smith's concept of humans and how an economy would self-organize has still not been fully formalized by mainstream economics or practically implemented.

However, this type of actor cannot thrive well in an institutional framework created for a *homo economicus*. The *homo socialis* requires different institutions. So, what institutions do we need in the 21st century?

In the past, societies have invented various institutions to overcome *tragedies of the commons*, for example, genetic favoritism, direct reciprocity ("I help you, if you help me"), or punitive institutions (such as police, courts, prisons and regulatory authorities). However, all of these mechanisms can lead to undesirable side effects such as ethnic conflicts or corruption. Our current system is built on punitive institutions, but such top-down regulation is very costly and creates many inefficiencies, including negative impacts on innovation.[16]

In Sect. 5.1, I will explain that, as systems get more and more complex, top-down regulation can no longer achieve efficient and satisfactory solutions. Instead, we can build on self-regulation, such as found in ecological, social and immune systems. I will illustrate the advantage of this approach by the example of urban traffic light control and will argue that a global reputation platform could become a suitable institution for our globalized world, the *global village* that we have created (see Sect. 6.2). This proposed approach supports trusted exchange, and increases opportunities for the participation of citizens in social, economic and political life. Furthermore, I will present the idea of a new, reputation-based kind of money, which I call *qualified money* (see Sect. 6.3). These new institutions would promote an *economy 2.0*, which will be a *participatory market society*, and help overcome the current gap between social and economic engagement (see Sect. 7).

My overall conclusion is that a change from an agent-oriented to an interaction-oriented view of decision-makers offers a better understanding of complex socio-economic systems, and leads to new solutions for long-standing problems. I will cite examples from the technology and business worlds, demonstrating that the transition to a *participatory market society* is already occurring. This is fueled by the digital revolution, particularly the Web 2.0 and social media platforms. If the right political decisions are taken pro-actively, this scenario can unleash the potential of many *networked minds* and lead to an age of creativity and sustainable prosperity.

---

16 Some of the leading industrial countries now have debts of around 100% of the GDP, which suggests that systems with many institutions for top-down regulation are very expensive and may not be sustainable in the long run.

2. A PARADIGM SHIFT IN ECONOMIC THINKING?

Like mathematics and physics, economics is proud of having an axiomatic foundation, and rightly so. In microeconomics, which focuses on the decision-making of actors such as individuals or firms, the paradigm of the self-regarding, optimizing agent prevails.[17] The market theory of supply, demand and equilibrium prices is based on Walras[18] and Arrow and Debreu[19]. In macroeconomics, there are historically two major competing schools: the Neoliberals who believe in such free markets[20] and the Keynesians who call for market regulation by politics.[21] As Lucas and Sargent[22] point out, the Keynesian view of economics lacks microeconomic foundations. Since then, New-Keynesianism provides a view of Keynesian economics with microeconomic foundations,[23] and other, somewhat hybrid views of macroeconomics have been proposed.[24] Nevertheless, the old debate of Neoliberalism versus (New-)Keynesianism keeps influencing much of public economic thinking and policy-making around the world, and there are still a number of unresolved issues:[25]

—

17 John von Neumann and Oskar Morgenstern, *Theory of Games and Economic Behavior* (New Jersey: Princeton University Press, 1944).

18 Léon Walras, *Éléments d'économie politique pure, ou Théorie de la richesse sociale* (in French) (Lausanne: L. Corbaz, 1877).

19 Arrow and Debreu, "Existence of an Equilibrium for a Competitive Economy".

20 Walras, Éléments d'économie politique pure; Alfred Marshall, *Principles of Economics* (Macmillan and Co, 1890); Milton Friedman, *Capitalism and Freedom* (University of Chicago: Chicago Press, 1962).

21 John Meynard Keynes, *The General Theory of Interest, Employment and Money* (Basingstoke: Palgrave Macmillan, 1936).

22 Robert Lucas and Thomas Sargent, "After Keynesian Macroeconomics," *Quarterly Review, Federal Reserve Bank of Minneapolis*, 3.2. (1979).

23 Nicholas G. Mankiw and David Romer, *New Keynesian Economics: Coordination Failures and Real Rigidities* (Vol. 2) (Massachusetts: MIT Press, 1991).

24 Bruce Greenwald and Joseph Stiglitz, "Keynesian, New Keynesian, and New Classical Economics," *Oxford Economic Papers* 39 (1987), p. 119–132.

25 Paul Krugman, "How did Economists Get it so Wrong?" *New York Times* (2009), http://www.nytimes.com/2009/09/06/magazine/06Economic-t.html?pagewanted=all&r=0; David Colander, M. Goldberg, A. Haas, K. Juselius, A. Kirman, T. Lux and B. Sloth, "The Financial Crisis and the Systemic Failure of the Economics Profession," *Critical Review* 21.2–3 (2009), p. 249–267; Alan Kirman "The Economic Crisis is a Crisis for Economic Theory," *CESifo Economic Studies* 56.4 (2010), p. 498–535; Andrew Haldane and Robert May, "Systemic Risk in Banking Ecosystems," *Nature* 469.(2011), p. 351–355; Thomas Lux and Frank Westerhoff, "Economics Crisis," *Nature Physics* 5.1 (2009), p. 2–3; Neil Johnson and Thomas Lux, "Financial Systems: Ecology and Economics," *Nature* 469.7330 (2011), p. 302–303; Paul Ormerod and Dirk Helbing, "Back to the Drawing Board for Macroeconomics," *What's the Use of Economics?: Teaching the Dismal Science after the Crisis*, ed. Diane Coyle (London: London Publishing Partnership, 2012); Dirk Helbing and Alan Kirman, "Rethinking Economics Using Complexity Theory," *Real-World Economics Review*, 64 (2013), p. 23–52; Dirk Helbing and Stefano Balietti (2010) "Fundamental and Real-World Challenges in Economics," *Science and Culture* 76.9–10 (2010), p. 399–417.

1. The two historic schools have at least partly incompatible views of economics.[26]

2. A consistent theoretical link between micro- and macroeconomics (which goes beyond the simplifying view of representative agent modeling) is lacking.[27]

3. Empirical and experimental findings are challenging some of the foundational assumptions of economics.[28]

4. The prevailing microeconomic view of decision-making is not easily compatible and integrated within the body of knowledge collected in anthropology, social psychology and sociology.[29]

5. The financial crisis has raised serious doubts that the most established economic theories can sufficiently predict, explain, or prevent financial meltdowns of markets and their impact on economies and societies.[30]

—

26 Greenwald and Stiglitz, "Keynesian, New Keynesian, and New Classical Economics;" Steve Keen, "Predicting the 'Global Financial Crisis': Post-Keynesian Macroeconomics," *The Economic Record* 89.285 (2013), p. 228–254.

27 Alan Kirman, "Whom or What Does the Representative Individual Represent?" *The Journal of Economic Perspectives* 6.2 (1992), p. 117–136.

28 Daniel Kahneman and Amos Tversky, "Prospect Theory: An Analysis of Decision under Risk," *Econometrica* 47.2 (1979), p. 263–291; Daniel Kahneman and Amos Tversky, "Choices, Values, and Frames," *The American Psychologist* 39.4 (1984), p. 341–350; Daniel Kahneman and Amos Tversky, "On the Reality of Cognitive Illusions," *Psychological Review* 103.3 (1996), p. 582–591, 592–596; Elizabeth Hoffman, Kevin McCabe and Vernon Smith, "Social Distance and Other-Regarding Behavior in Dictator Games," *The American Economic Review* 86 (1996): 653–660; James Hartley, *The Representative Agent in Macroeconomics* (London: Routledge, 1997); Amos Tversky and Daniel Kahneman, "Advances in Prospect Theory: Cumulative Representation of Uncertainty," *Journal of Risk and Uncertainty* 5.4 (1992), p. 297–323; Reinhard Selten and Axel Ockenfels, "An Experimental Solidarity Game," *Journal of Economic Behavior & Organization* 34.4 (1998), p. 517–539; Ernst Fehr and Klaus Schmidt, "A Theory of Fairness, Competition, and Cooperation," *The Quarterly Journal of Economics* 114.3 (1999), p. 817–868; Ernst Fehr and Urs Fischbacher, "The Nature of Human Altruism," *Nature* 425.6960 (2003), p. 785–791; Dan Ariely, *Predictably Irrational: The Hidden Forces that Shape Our Decisions* (New York: HarperCollins Publishers, 2009).

29 Siegwart Lindenberg, "Homo socio-oeconomicus: The Emergence of a General Model of Man in the Social Sciences," (1990); Herbert Gintis, *The Bounds of Reason: Game Theory and the Unification of the Behavioral Sciences* (Princeton: Princeton University Press, 2009); Daniel McFadden, "The New Science of Pleasure. Consumer Choice Behavior and the Measurement of Well-Being," (2012) *preprint, see* http://emlab.berkeley.edu/wp/mcfadden122812.pdf.

30 Colander et al., "The Financial Crisis and the Systemic Failure of the Economics Profession;" Kirman, "The Economic Crisis is a Crisis for Economic Theory;" Steve Keen, *Debunking Economics. Revised and Expanded Edition: The Naked Emperor Dethroned?* (London: Zed Books, 2011); Hans Gersbach and Jean-Charles Rochet, "Aggregate Investment Externalities and Macro Prudential Regulation" (*Journal of Money, Credit and Banking* 44 Issue Supplements 2) (2012), p. 73–109; Mathias Dewatripont and Jean Tirole "Macroeconomic Shocks and Banking Regulation" (*Journal of Money, Credit and Banking* 44 Issue Supplements 2) (2012), p. 237–254; Keen, "Predicting the 'Global Financial Crisis': Post-Keynesian Macroeconomics".

Are we perhaps facing economic problems, because our theoretical picture of economies does not fit reality well enough? To answer this, scientists need to question established knowledge. In fact, many new approaches have been proposed in the past decades. Due to spatial limitations I can mention only a few. For example, I would refer to the work of Brian Arthur on innovation,[31] of Reinhard Selten in experimental economics,[32] of Joseph Stiglitz on economic downturns,[33] of Paul Krugman on the role of geography,[34] of Daniel Kahnemann and Amos Tversky on risk perception,[35] of George Akerlof and Robert Shiller on herding behavior (*animal spirits*, see Akerlof and Shiller),[36] of Gerd Gigerenzer on heuristics,[37] of Bruno Frey on happiness,[38] of Ernst Fehr on fairness,[39] of Herbert Gintis on evolutionary game theory,[40] of Ulrich Witt on evolutionary economics,[41] and of Alan Kirman and Mauro Gallegati on the role of heterogeneity and non-representative agents in economics.[42] I would also mention work presenting a complexity science perspective on economic systems[43] or an econophysics

---

31 Brian Arthur, "Competing Technologies, Increasing Returns, and Lock-In by Historical Events," *The Economic Journal* 99.394 (1989), p. 116–131; Brian Arthur, "Complexity and the Economy," *Science* 284.5411 (199), p. 107–109.

32 Selten and Ockenfels, "An Experimental Solidarity Game".

33 Joseph Stiglitz, "Rethinking Macroeconomics: What Failed, and How to Repair it," *Journal of the European Economic Association* 9.4 (2011), p. 591–645.

34 Paul Krugman, "Increasing Returns and Economic Geography," *Journal of Political Economy* 99.3 1991, p. 483–499.

35 Kahneman and Tversky, "Prospect Theory: An Analysis of Decision under Risk;" Kahneman and Tversky, "Choices, Values, and Frames;" Tversky and Kahneman, "Advances in Prospect Theory: Cumulative Representation of Uncertainty;" Kahneman and Tversky, "On the Reality of Cognitive Illusions".

36 George Akerlof and Robert Shiller, *How Human Psychology Drives the Economy, and Why It Matters for Global Capitalism* (Princeton: Princeton University Press, 2010).

37 Gerd Gigerenzer and Peter Todd, ABC Research Group *Simple Heuristics that Make Us Smart* (Oxford: Oxford University, 1999).

38 Bruno Frey, *Happiness: A Revolution in Economics* (Boston, Massachusetts: MIT Press, 2008).

39 Fehr and Schmidt, "A Theory of Fairness, Competition, and Cooperation;" Fehr and Fischbacher, "The Nature of Human Altruism".

40 Herbert Gintis, *Game Theory Evolving: A Problem-Centered Introduction to Modeling Strategic Interaction* (Princeton: Princeton University Press, 2009).

41 Ulrich Witt, "Imagination and Leadership – the Neglected Dimension of an Evolutionary Theory of the Firm," *Journal of Economic Behavior & Organization* 35.2 (1998), p. 161–177.

42 Mauro Gallegati and Alan Kirman, *Beyond the Representative Agent* (Cheltenham: Edward Elgar, 1999); Alan Kirman and Jean-Benoit Zimmermann, *Economics with Heterogeneous Interacting Agents* (New York: Springer, 2001); Alan Kirman, "Heterogeneity in Economics," *Journal of Economic Interaction and Coordination* 1.1 (2006), p. 89–117; Domenico Delli Gatti, E. Gaffeo, M. Gallegati, G. Giulioni, and A. Palestrini, *Emergent Macroeconomics: An Agent-based Approach to Business Fluctuations* (New York: Springer, 2008).

43 Paul Krugman, *The Self-Organizing Economy* (New York: Blackwell Publishers, 1996); William Brock and Cars Hommes, "A Rational Route to Randomness," *Econometrica* 65.5 (1997): 1059–1095; Richard Day, *Complex*

perspective.[44] So far, however, a unifying theoretical framework of all these important findings and their derivation from first principles is largely lacking, in contrast to the results of mainstream economics. Perhaps for this reason, many foundational economics courses still suggest that markets can be understood *as if* people behaved according to the idealized assumptions of mainstream economics.[45] It, therefore, seems that most economic work still views man as an – in many ways – imperfect approximation of a *homo economicus*, and that these imperfections would more or less average out on the macro-level.

In response to the low predictive power of many economic theories and the economic problems most countries face today, it is often argued that mainstream economic principles are idealizations rather than a faithful representation of the real world. This calls for a better understanding of the deviations between currently established theories on the one hand, and empirical or experimental evidence on the other. Perhaps it even requires a fundamentally new way of thinking.

In fact, many eminent thinkers call for a different approach. They include Nobel prize laureates and ex-presidents of central banks. The most explicit statements are probably those of Paul Krugman, who in 2009 asked: "How did economists get it so wrong?," and Alan Greenspan, who acknowledged in testimony before a congressional committee on October 23, 2008, that "the whole intellectual edifice [ ] collapsed in the summer of last year".[46] Highlighting the contradictions in economics, Krugman has since famously

---

Economic Dynamics: An Introduction to Macroeconomic Dynamics, Vol. 2 (Massachusetts: MIT Press, 1999); Sunny Auyang, *Foundations of Complex-System Theories: In Economics, Evolutionary Biology, and Statistical Physics* (Cambridge: Cambridge University Press, 1999); Thomas Lux and Michele Marchesi, "Scaling and Criticality in a Stochastic Multi-Agent Model of a Financial Market," *Nature* 397.6719 (1999), p. 498–500; Marisa Faggini and Thomas Lux, *Coping with the Complexity of Economics* (New York: Springer, 2009); Alan Kirman, *Complex Economics: Individual and Collective Rationality* (London: Routledge, 2010); Helbing and Balietti, "Fundamental and Real-World Challenges in Economics;" Keen, "Predicting the 'Global Financial Crisis'".

44 Jean-Philippe Bouchaud and Marc Potters, *Theory of Financial Risks: From Statistical Physics to Risk Management* (Cambridge: Cambridge University Press, 2000); Didier Sornette, *Why Stock Markets Crash: Critical Events in Complex Financial Systems* (Princeton: Princeton University Press, 2004); Masanao Aoki and Hiroshi Yoshikawa, *Reconstructing Macroeconomics: A Perspective from Statistical Physics and Combinatorial Stochastic Processes* (Cambridge: Cambridge University Press, 2006); Bikas Chakrabarti, Anirban Chakraborti and Arnab Chatterjee, *Econophysics and Sociophysics* (Weinheim: Wiley-VCH, 2006); Arnab Chatterjee and Bikas Chakrabarti, *Econophysics of Markets and Business Networks* (New York: Springer, 2007); Sitabhra Sinha, A. Chatterjee, A. Chakraborti and B. Chakrabarti, *Econophysics: An Introduction* (Weinheim: Wiley-VCH, 2011).

45 The book by Paul Bowles, R. Edwards and F. Roosevelt, *Understanding Capitalism* (Oxford: Oxford University Press, 2005) appears to be an exception.

46 For more quotes see Helbing and Balietti, "Fundamental and real-world challenges in economics," and Helbing and Kirman, "Rethinking Economics Using Complexity Theory".

attempted to defend the existing micro-economic toolkit in a debate with Keen,[47] and Greenspan has been criticized for his continued use of these tools that he acknowledges have been disproved. It is therefore not surprising that most financial traders do not seem to apply or believe in scholarly economics. Given the huge confusion, with economists using models they deem unsuitable and professionals either using possibly flawed tools or not using established economic theories at all, it is apparent that the economic foundations should be seriously reconsidered. Indeed, George Soros has established an *Institute of New Economic Thinking*,[48] and Edward Fullbrock and others have founded the *World Economics Association* to support heterodox economics with journals such as *Real-World Economics Review*.[49] Certainly, this is only the beginning.

While theories in microeconomics are mostly based on the concept of the *homo economicus*, it must also be recognized that identical macro-level phenomena might be explained by different kinds of micro-level assumptions, as much relevant information is lost through macro-level aggregation.[50] There are alternative micro-level theories which are worth considering.[51] For example, rather than taking the *homo economicus* as a theoretical starting point, it would be equally justified to start from the *homo socialis*, i.e. decision-makers with other-regarding preferences (see Sec. 3).

While there is empirical evidence for other-regarding behavior[52] and some theoretical implications of this have been studied,[53] a good theoretical foundation of the *homo socialis* has long been lacking. Recent theoretical progress,[54] however, suggests that a theory for the *homo socialis* can be rigorously derived from first principles, rooted in the best response rule (utility maximization) and principles of evolution (see Sec. 3). This

---

47 Keen, "Predicting the 'Global Financial Crisis': Post-Keynesian Macroeconomics".

48 See http://ineteconomics.org.

49 See http://www.worldeconomicsassociation.org and http://www.paecon.net/PAEReview.

50 Dirk Helbing, "Pluralistic modeling of complex systems," *Science and Culture* 76.9–10 (2010), p. 315–329.

51 Gintis, *The Bounds of Reason: Game Theory and the Unification of the Behavioral Sciences*.

52 Werner Güth, R. Schmittberger and B. Schwarze (1982) "An Experimental Analysis of Ultimatum Bargaining," *Journal of Economic Behavior & Organization* 3.4 (1982), p. 367–388; Joe Henrich, R. Boyd, S. Bowles, C. Camerer, E. Fehr, H. Gintis and R. McElreath, "In Search of Homo Economicus: Behavioral Experiments in 15 Small-Scale Societies," *The American Economic Review* 91.2 (2001), p. 73–78. Fehr and Fischbacher, "The Nature of Human Altruism;" Bruno Frey, *Happiness: A Revolution in Economics* (Boston, Massachusetts: MIT Press, 2008).

53 Martin Dufwenberg, P. Heidhues, G. Kirchsteiger, F. Riedel and J. Sobel, "Other-Regarding Preferences in General Equilibrium," *The Review of Economic Studies* 78.2 (2011), p. 613–639; Joel Sobel, "Generous Actors, Selfish Actions: Markets with Other-Regarding Preferences," *International Review of Economics* 56 (2009), p. 3–16.

54 Thomas Grund, C. Waloszek and D. Helbing, "How Natural Selection Can Create Both Self-and Other-Regarding Preferences, and Networked Minds," *Scientific Reports* 3 (2013), p. 1480.

raises the pertinent question whether such a theory might better explain empirical and experimental evidence, and perhaps create a paradigm shift in economic thinking in due course.[55]

For example, the fact that the *homo socialis* is influenced by the interests and preferences of others explains a mysterious, but widely observed fact in sociology, namely social influence.[56] Social influence helps to understand typical characteristics of opinion formation[57] and herding effects (*animal spirits*), including bubbles and crashes at financial markets.[58]

I would also expect that the incorporation of the theory of the *homo socialis* will provide new insights and a fundamentally new understanding of social capital, power, reputation, trust as well as economic value,[59] as these quantities result from social network interactions. Such network interactions are characteristic for the *homo socialis*, who is perhaps best characterized by conditional cooperativeness and the term *networked minds* (see below). I expect that a theory of the *homo socialis* will be able to consistently explain many *puzzles* in the social sciences and economics. Currently, we still know little about the *homo socialis*, but what we do know is very interesting. Therefore, there is much exciting research to be done. I argue that the *homo socialis* would justify a new branch of economics that one might call *economics 2.0* or *socionomics* (see Sec. 5).[60] So, what exactly is the *homo socialis* about?

—

55 Empirical evidence suggests that more than 60% of subjects have other-regarding preferences (Ryan Murphy, K. A. Ackermann and M. J. J. Handgraaf, "Measuring Social Value Orientation," *Judgment and Decision Making* 6.8 (2011): p . 771–781) but the percentage may also depend on the socio-cultural background.

56 Solomon Asch, "Effects of Group Pressure upon the Modification and Distortion of Judgments," *Documents of Gestalt Psychology*, ed. Mary Henle (Berkeley: University of California Press, 1951), p. 222–236; Solomon Asch, "Studies of Independence and Conformity: I. A Minority of One against a Unanimous Majority," *Psychological Monographs* 70.9 (1956), p. 1–70.

57 Michael Mäs, A. Flache and D. Helbing, "Individualization as Driving Force of Clustering Phenomena in Humans," *JPLoS Computational Biology*, 6.10: e1000959 (2010).

58 Vernon L. Smith, G. L. Suchanek and A. W. Williams (1988) "Bubbles, Crashes, and Endogenous Expectations in Experimental Spot Asset Markets," *Econometrica* 56.5 (1988), p. 1119–1151; Gunduz Caginalp, D. Porter and V. Smith, "Financial Bubbles: Excess Cash, Momentum, and Incomplete Information," *The Journal of Psychology and Financial Markets* 2.2 (2001), p. 80–99; Sornette, *Why Stock Markets Crash*; Cars Hommes, J. Sonnemans, J. Tuinstra and H. van de Velden, "Coordination of Expectations in Asset Pricing Experiments," *Review of Financial Studies* 18.3 (2005), p. 955–980; Cars Hommes, J. Sonnemans, J. Tuinstra and H. van de Velden, "Expectations and Bubbles in Asset Pricing Experiments," *Journal of Economic Behavior & Organization* 67.1 (2008), p. 116–133; Akerlof and Shiller, *How Human Psychology Drives the Economy.*

59 Nan Lin, *Social Capital* (London: Routledge, 2010).

60 For a related video lecture see http://www.youtube.com/watch?v=Ef2Ag—rwouo. Note that the methodological approach and scientific understanding proposed in this paper has little overlap with the current, mood-focused approach by Robert Prechter, D. Goel, W. D. Parker and M. Lampert, "Social Mood,

3. THE EMERGENCE OF THE *HOMO SOCIALIS*

Establishing a new economic thinking is not just obstructed by a lack of alternative (*heterodox*) models. If one departs from the concept of the *homo economicus*, i.e. the strictly optimizing self-regarding agent, there are myriads of possibilities how agents and their decision-making rules may be defined. However, no alternative model stands out thus far, and a consensus on a non-homo-economicus-kind of model has not emerged. In some sense, there is too much arbitrariness in specifying such models. But this might just have changed.

3.1 Utility-maximizing agents under evolutionary pressure

A recently published paper[61] studied whether evolution has made humans *nasty* (self-regarding) or *nice* (other-regarding).[62] It analyzes social dilemma situations for the typical example of the *Prisoner's Dilemma game* (PD), which distinguishes two kinds of behavior: *cooperation* and *defection*. The PD takes into account the risk of cooperation and the temptation to defect.

For Prisoner's Dilemma interactions, cooperative behavior is irrational for a *homo economicus*, resulting in a *tragedy of the commons* with poor payoffs.[63] To stabilize cooperation, social mechanisms such as direct, indirect, spatial or network reciprocity, group competition, or costly punishment have been proposed, often assuming imitation of better-performing behaviors.[64] These mechanisms, like taxes, effectively transform the Prisoner's Dilemma into other kinds of games that are more favorable for cooperation.[65] So far, however, other-regarding preferences[66] lacked a convincing explanation.[67]

—

Stock Market Performance and U.S. Presidential Elections: A Socionomic Perspective on Voting Results," (September 27, 2012), see http://www.socionomics.net/.

61 Grund et al., "How Natural Selection Can Create Both Self-and Other-Regarding Preferences, and Networked Minds".

62 See also the related article in the *Wall Street Journal* of March 29, 2013: http://online.wsj.com/article/SB10001424127887324105204578384930047065520.html.

63 Garrett Hardin, "The Tragedy of the Commons," *Science* 162.3859 (1968), p. 1243–1248.

64 Karl Sigmund, *Games of Life: Explorations in Ecology, Evolution, and Behaviour* (USA: Penguin Books, 1995); Fehr and Fischbacher, "The Nature of Human Altruism;" Martin A. Nowak, *Evolutionary Dynamics: Exploring the Equations of Life* (Cambridge, Massachusetts: Harvard University Press, 2006); Gintis, *Game Theory Evolving*.

65 Dirk Helbing and Sergi Lozano, "Phase Transitions to Cooperation in the Prisoner's Dilemma," *Physical Review* E 81.5 (2010): 057102.

66 Not to be confused with other-regarding behavior, which is the same as cooperation.

67 I.e. an explanation, which does not assume favorable but debatable mechanisms such as genetic drift or cultural selection (in the sense of a social modification of the reproduction rate).

The above-mentioned study[68] distinguishes individual preferences from behavior. It investigates Prisoner's Dilemma interactions in two-dimensional space on the basis of very few assumptions, each of which promotes self-regarding preferences or defection:

1. Agents decide according to a best-response rule that strictly maximizes their utility function, given the behaviors of their interaction partners (their neighbors).

2. The utility function considers not only the own payoff, but gives a certain weight to the payoff of their interaction partner(s). The weight is called the *friendliness* and set to zero for everyone at the beginning of the simulation.

3. Friendliness is a trait that is inherited (either genetically or by education) to offspring. The likelihood to have offspring increases exclusively with the own payoff, not the own utility. The payoff is assumed to be zero, when a friendly agent is exploited by all neighbors (i.e. if they all defect). Therefore, such agents will never have any offspring.

4. The inherited friendliness value tends to be that of the parent. There is also a certain mutation rate, but it does not promote friendliness. (In the simulation results discussed here, mutations were specified such that they imply an average friendliness of 0.2, which cannot explain the typically observed value of 0.4).

Based on the above assumptions, the characteristic outcome of the evolutionary game-theoretical computer simulations is a self-regarding, payoff-maximizing *homo economicus*, as expected. But this applies only to *most* parameter combinations. When offspring tend to live close to their parents (i.e. intergenerational migration is low), a friendly *homo socialis* with other-regarding preferences results instead, see Fig. 1.[69] This is quite surprising, since the above assumptions do not favor such an outcome. None of these rules promotes other-regarding preferences or cooperation in separation (i.e. they might be considered socially dysfunctional), but they are nevertheless creating socially favorable behavior in combination. This can only be explained as the result of an interaction effect between the above rules. Another interesting finding is the evolution of *cooperation between strangers* (see Fig. 2), i.e. the emergence of the *homo socialis* does not require genetic favoritism.

---

68 Grund et al., "How Natural Selection Can Create Both Self-and Other-Regarding Preferences, and Networked Minds".

69 The *homo socialis* may be defined as being sensitive to the social context, while the "homo economicus" is not. However, the other-regarding preference must not be reflected by a utility function that weights the payoffs of others, see Sect. 5.1.

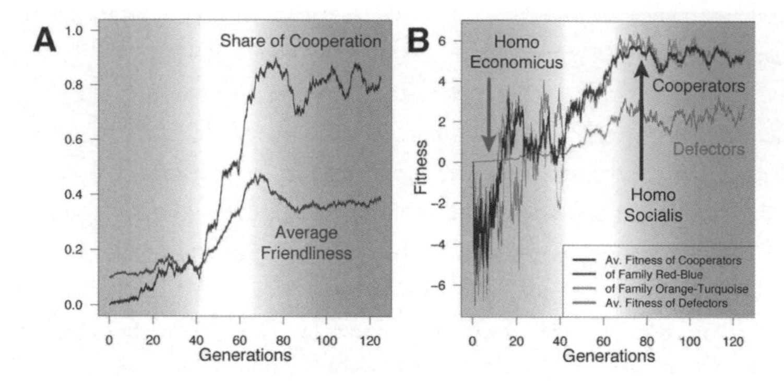

Fig. 1. Emergence of the *homo socialis* in a world initially dominated by the *homo economicus*. (A) Average value of the friendliness and of the proportion of cooperating agents as a function of time. One generation corresponds to $1/\beta$ update time steps, where β is the death rate. (B) Average payoffs of cooperators and defectors as a function of time. At the beginning, defectors are more successful than cooperators, as they receive higher payoffs on average. However, after many generations, a transition from a *homo economicus* to a *homo socialis* occurs. Then, the payoffs for cooperating agents (which are of the *homo socialis* type) are higher than the payoffs for defectors. This allows the *homo socialis* to spread thanks to a reproduction rate that is now higher than that of the *homo economicus*. Note that the population dynamics implies that families might die out before high levels of cooperation are reached. For details of the underlying simulations see the caption of Fig. 2 and Grund et al., "How Natural Selection Can Create Both Self-and Other-Regarding Preferences, and Networked Minds".

How can we understand this outcome? Mutations generate a certain level of friendliness by chance (*mutations*). This slight other-regarding preference (eventually giving the payoff of others an average weight of about 0.2) creates conditionally cooperative behavior, as postulated by Fehr and others.[70] That is, if enough neighbors cooperate, a *conditional cooperator* will be cooperative as well, but not so if too many neighbors defect.[71] Unconditionally cooperative agents with a high level of friendliness are born very rarely,

---

70 Fehr and Schmidt, "A Theory of Fairness, Competition, and Cooperation".

71 Conditional cooperators can exist for friendliness values ρi with $(T-R)/(T-S)<\rho i<(P-S)/(T-S)$ (Grund et al., "How Natural Selection Can Create Both Self-and Other-Regarding Preferences, and Networked Minds"), with the payoff values P, R, S, T defined in the caption of Fig. 2. Below the lower threshold, an agent always defects and corresponds to a *homo economicus*. Above the higher threshold, an agent is unconditionally cooperative (an idealist). As a consequence, if all agents would have an identical friendliness ρ, one would expect a hysteretic system behavior as a function of the friendliness value. When the friendliness increases from low to high values, a discontinuous transition from defection to cooperation of everyone is expected when the value $(P-S)/(T-S)$ is exceeded, while a discontinuous transition from a cooperative to a defective behavior of everyone is expected when the friendliness value becomes smaller than $(T-R)/(T-S)$.

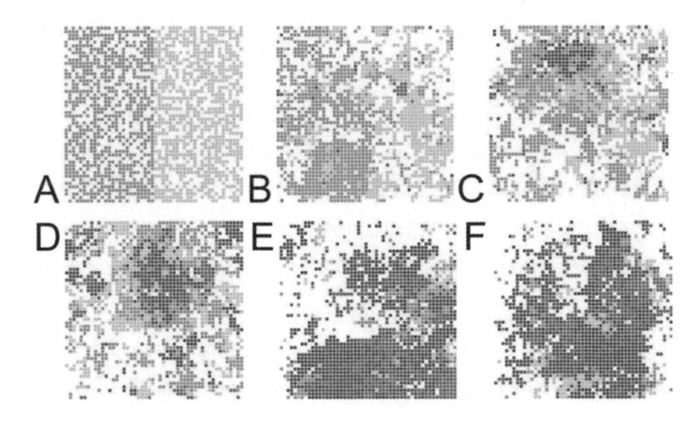

Fig. 2. Evolution of *cooperation between strangers* (here: cooperation between two different families). The above figure shows snapshots (at times $t=0$, 250, 500, 1000, 2000, and $t=2500=125$ generations) of a representative simulation run for two families (i.e. two kinds of genetically related individuals), populating 60% of a 50×50 spatial grid without periodic boundary conditions (For a related video see http://vimeo.com/65376719). The initial friendliness in the light-grey family is 0, and in the middle-grey-black family it is assumed to be 0.2 (but it could be chosen 0 as well). (Grund et al., "How Natural Selection Can Create Both Self-and Other-Regarding Preferences, and Networked Minds".) One finds the emergence of cooperation in both families (represented by black and grey), which cannot occur for the *homo economicus*. Hence, cooperation implies a *homo socialis*. Remarkably, there are clusters in which members of different families cooperate (i.e. a *cooperation between strangers* evolves). Individuals play 2-person prisoner's dilemma games with all neighbors in their respective Moore neighborhoods (but similar results are expected for some other social dilemma games as well). The payoff parameters are the *Reward* $R=1$ for the cooperation of both interaction partners, the *Punishment* $P=0$ for non-cooperative behavior on both sides, the *Temptation* $T=1.1$ for a defector exploiting a cooperator, and the *Sucker's Payoff* $S=-1$ for an exploited cooperator. Each agent's utility function weights the payoffs of the neighbors with the *friendliness* ρ_i, i.e. the utility function U_i of an agent i is assumed to be $U_i=(1-\rho_i)P_i+\rho_i\langle P\rangle_i$, where P_i is the payoff of individual i and $\langle P\rangle_i$ the average payoff of the neighbors. For all agents, the initial friendliness is set to $\rho_i=0$ (corresponding to a *homo economicus*) and the initial behavior (*strategy*) is assumed to be defective (light-grey or middle grey). The behaviors of all individuals are simultaneously updated based on a best response rule, which implies a behavior that maximizes the individual utility based on the behaviors of the neighbors. Only with a probability of 0.05 do individuals take other decisions than the best response rule suggests. To allow for evolution, while keeping population size constant, individuals die at random with the probability $\beta=0.05$, but are replaced with offspring of living individuals. The likelihood to give birth to offspring is proportional to the actual payoffs in the previous round (i.e. not proportional to the utility). With probability ν, the offspring is born in the closest empty site to the parent, while with probability $(1-\nu)$, the offspring moves to a randomly selected empty site (here: $\nu=0.95$). Offspring inherit a trait called friendliness ρ_i from their parents, which is subject to random mutations. (With probability 0.8 the offspring's friendliness is *reset* to a uniformly distributed random value between 0 and the friendliness ρ_i of the parent, and with probability 0.2 it takes on a uniformly distributed value between ρ_i and 1.) The local reproduction rate ν determines, whether a transition from a *homo economicus* with self-regarding preferences to a *homo socialis* with other-regarding preferences takes place. (See Fig. 2 in Grund et al., "How Natural Selection Can Create Both Self-and Other-Regarding Preferences, and Networked Minds".)

and only by chance. These *idealistic* individuals will usually be exploited, have very poor payoffs, and no offspring. However, if born into an environment where enough agents have a moderate friendliness and are conditionally cooperative, *idealists* with a high level of friendliness and unconditionally cooperative behavior can trigger cooperative behavior of the neighbors in a cascade-like manner. Under such conditions, high levels of friendliness are passed on to many offspring such that a *homo socialis* with other-regarding preferences emerges and spreads. This holds, because greater friendliness now tends to be profitable. On average, now cooperators earn higher payoffs than defectors. Hence, the *homo socialis* can eventually outcompete the *homo economicus*, while initially the *homo economicus* earns more (see Fig. 1). Nevertheless, the friendliness levels are widely distributed, thereby explaining heterogeneous individual preferences, as empirically observed.[72] In other words, everything from selfish to altruistic preferences exists.

In the situation studied above, where everyone starts as a defecting *homo economicus*, no single individual can establish profitable cooperation, not even by optimizing decisions over an infinitely long time horizon. It takes a few *friendly* deviations to trigger cascade effects that eventually change the macro-level outcome. Therefore, it is important to recognize that a critical number of interacting individuals needs to be friendly and cooperative by coincidence. The *homo socialis* would never evolve without the occurrence of random *mistakes* (here: the birth of *idealists* who are exploited by everyone). However, given suitable feedback effects, such *errors* enable better systemic outcomes. Here, they eventually overcome the *tragedy of the commons*,[73] resulting in an *upward spiral* towards cooperation with high payoffs. Remarkably, this is not the outcome of an optimization process, but rather of an evolutionary process.

The simple evolutionary theory described in the paper discussed above contains only a few fundamental and widely accepted assumptions, and might have the potential to form the basis of an integrated theoretical approach in the behavioral sciences. In accordance with a lot of work in social psychology, the *homo socialis* may be characterized by *empathy* in the sense that the *homo socialis* puts himself or herself into the shoes of others. More precisely, the *homo socialis* takes into account the perspective, interests, and success of others when taking own decisions.[74] This effectively explains the fairness

---

72 Ryan O. Murphy, K. A. Ackermann and M. J. J. Handgraaf, "Measuring Social Value Orientation," *Judgment and Decision Making* 6.8 (2011), p. 771–781.

73 Hardin, "The Tragedy of the Commons".

74 However, taking account the interests of others must not necessarily be based on a weighted utility function, see the example on traffic light control in Sect. 5.1.

preferences observed in behavioral economics.[75] Note, however, that the *homo socialis* should not be imagined as a *homo economicus* sharing some payoff with others. The *homo socialis* decides differently from the *homo economicus*, and that is why *tragedies of the commons* can be overcome.[76]

Decisions of the *homo socialis* are interdependent, in contrast to the *homo economicus*, who takes decisions indendently without the consideration of effects on others. Therefore, the *homo socialis* may be characterized by the term *networked minds*. This implies a complex dynamics of systems where many such decision-makers interact, with important economic implications.[77] While methods from statistics should be good enough to characterize the *homo economicus*, the description of the *homo socialis* requires methods drawn from complexity science.

3.2 The FuturICT initiative

When non-linear interactions and network interdependencies prevail, economic systems are expected to be complex dynamical systems.[78] From ecology, climate research, and statistical physics it is known that such systems cannot be well understood merely by analytical and econometric approaches, in particular because of emergent system properties resulting from interactions between the system components.[79]

A characteristic feature of complex dynamical systems is that they may not necessarily be in equilibrium,[80] and that a representative agent approach may not work

---

75 Güth et al., "An Experimental Analysis of Ultimatum Bargaining;" Fehr and Schmidt, "A Theory of Fairness, Competition, and Cooperation;" Henrich et al., "In Search of Homo Economicus: Behavioral Experiments in 15 Small-Scale Societies;" Fehr and Fischbacher, "The Nature of Human Altruism".

76 One might say the *homo socialis* makes the individual and system optimum (more) compatible with each other.

77 Helbing and Kirman, "Rethinking Economics Using Complexity Theory".

78 Krugman, *The Self-Organizing Economy*; Brock and Hommes, "A Rational Route to Randomness;" Day, *Complex Economic Dynamics: An Introduction to Macroeconomic Dynamics*; Lux and Marchesi, "Scaling and Criticality in a Stochastic Multi-Agent Model of a Financial Market;" Auyang, *Foundations of Complex-System Theories*; Faggini and Lux, *Coping with the Complexity of Economics*; Kirman, Complex Economics, Helbing and Balietti, "Fundamental and Real-World Challenges in Economics;" Keen, "Predicting the 'Global Financial Crisis';" Brian Arthur, "Complexity Economics: A Different Framework for Economic Thought," *Complexity Economics*, ed. Brian Arthur (Oxford: Oxford University Press, 2013) (forthcoming).

79 Steven Strogatz, *Nonlinear Dynamics and Chaos* (New York Perseus Books, 1994)

80 Dirk Helbing, S. Lämmer, U. Witt, and T. Brenner, "Network-Induced Oscillatory Behavior in Material Flow networks and Irregular Business Cycles," *Physical Review* E 70 (2004): 5, 056118; Gatti et al., *Emergent Macroeconomics*.

as well,[81] partly because of strong correlations and cascade effects.[82] Therefore, the FuturICT initiative[83] has recently proposed to combine socio-economic approaches with complexity science, massive computer power and Big Data to analyze, model and simulate techno-socio-economic-environmental systems. The initiative aims to develop a better understanding of the highly interdependent and densely networked world we are living in, to increase its sustainability and resilience.

The FuturICT initiative points out that the globalization and the increasing connectedness in human-influenced (*anthropogenic*) systems has increased the complexity of our world and also created systemic risks.[84] It is argued that, due to emergent phenomena, densely connected and strongly interacting systems cannot be understood from the properties of the components and the behaviors of separately deciding agents.[85] This implies the need to fundamentally change our conceptional approach, namely to move from a component-oriented perspective to an interaction-oriented, systemic perspective. Consequently, policy implications may look quite different from what the theory of independent decision-makers suggests, but this may help to overcome some long-standing problems.[86]

In order to ensure that our knowledge can keep up with the speed of change of our complex, globalized world, the FuturICT initiative also proposes to establish a "Global Systems Science" aiming at a "grand integration of knowledge" by combining the best knowledge of the social, natural and engineering sciences in a large-scale and truly interdisciplinary effort.[87] Despite competition, I believe that scientific curiosity can

---

81 Gallegati and Kirman, *Beyond the Representative Agent*; Hiroshi Deguchi, *Economics as an Agent-based Complex System* (New York: Springer, 2004); Dirk Helbing, *Social Self-Organization: Agent-based Simulations and Experiments To Study Emergent Social Behavior* (New York: Springer, 2012).

82 Stefano Battiston, D. Delli Gatti, M. Gallegati, B. Greenwald and J. Stiglitz, "Liaisons dangereuses: Increasing Connectivity, Risk Sharing, and Systemic Risk," *Journal of Economic Dynamics & Control* 36.8 (2012): p. 1121–1141; Gabriele Tedeschi, A. Mazloumian, M. Gallegati and D. Helbing, "Bankruptcy Cascades in Interbank Markets," *PLoS ONE* 7.12 (2012): e52749; Helbing, "Globally Networked Risks and How to Respond".

83 See http://www.futurict.eu.

84 Helbing, "Globally Networked Risks and How to Respond".

85 Dirk Helbing and Anna Carbone, "Participatory Science and Computing for our Complex World," *EPJ Special Topics* 214 (2012), p. 1–666.

86 Massimo Salzano and David Colander, *Complexity Hints for Economic Policy* (New York: Springer, 2007); Helbing, *Social Self-Organization*; Helbing and Kirman, "Rethinking Economics Using Complexity Theory".

87 Helbing, "Globally Networked Risks and How to Respond".

promote a respectful interaction and cooperation that can generate insights on all sides. For this, one should promote a culture of cross-disciplinary appreciation.

4. DIFFERENCES BETWEEN THE *HOMO SOCIALIS* AND THE *HOMO ECONOMICUS*

One may argue that the *homo socialis*, who considers the payoff of others, has been covered by rational choice theory already,[88] since it has been increasingly recognized that (many) people do not just optimize their own payoffs (as shown, for example, by Ultimatum Game experiments).[89] To take this into account, it is assumed that individual preferences are part of the utility function and that altruistically behaving individuals enjoy helping others, while the others do not. However, when the utility function is treated as an individual rather than a universal quantity, a rational choice theory based on utility maximization loses much of its predictability and strength. Then, the utility function cannot be calculated from first principles, but it must statistically be fitted to data separately for each individual. In contrast, the above theory of the *homo socialis* provides an evolutionary explanation of the individual utility function, which is a major advantage.

In fact, both, the *homo economicus* and the *homo socialis* assume self-determined individual decisions, according to different utility functions. At first glance, it may appear that this difference is not important. However, when the *friendliness* parameter in the utility function is changed continuously, one eventually crosses a *tipping point* beyond which the resulting individual and system behavior look dramatically different. In other words, when the friendliness parameter is varied, the system behavior changes discontinuously. Individual optimization attempts will only simultaneously create individual prosperity and maximum social benefits beyond a certain level of *friendliness*, i.e. for the *homo socialis*.

In other words, in social dilemma situations, Adam Smith's principle of the *invisible hand* works only for the *homo socialis*, but not for the *homo economicus*. Without suitable

---

88 Theodore Bergstrom, "Systems of Benevolent Utility Functions," *Journal of Public Economic Theory* 1.1 (1999), p. 71–100; Yann Bramoullé, "Interdependent Utilities, Preference Indeterminacy, and Social Networks," *Working paper*, http://www.eea-esem.com/papers/eea-esem/eea2002/1618/interdependent-utilities.pdf (2001).

89 See Güth et al., "An Experimental Analysis of Ultimatum Bargaining;" Henrich et al., "In Search of Homo Economicus: Behavioral Experiments in 15 Small-Scale Societies".

institutional settings or regulations, the *homo economicus* will run into a *tragedy of the commons* with very poor payoffs for the great majority. Note that the principle of the *invisible hand* would work for the *homo economicus* as well, if the underlying assumptions of the First Theorem of Welfare Economics were fulfilled, i.e. if all individuals would interact in one shared market without transaction costs and externalities.[90] However, these assumptions are quite restrictive, and the macro-level outcome depends on them in a sensitive manner. For example, non-centralized network interactions and transaction costs can manifestly change the outcome.[91]

As the *homo economicus* decides strictly according to self-interest, the coordination of decision-makers and efforts to overcome *tragedies of the commons* or market failures requires some kind of top-down regulation. This results in a steadily growing number of laws and regulations and costly investments in regulatory and punitive institutions (such as police, courts, market regulators, etc.). This can significantly decrease systemic efficiency. In addition, regulation tends to reduce diversity, which can affect innovation in a negative way.[92]

In comparison with the *homo economicus*, the *homo socialis* is able to cope with social dilemma situations based on local interactions. I will characterize this as bottom-up *self-regulation*, considering that the emergent level of *friendliness* overcoming the *tragedy of the commons* results from a decentralized, evolutionary process. Since top-down regulation (as required for the *homo economicus*) is just the opposite of bottom-up self-regulation (as typical for the *homo socialis*), it is misleading to say that the *homo socialis* is just a variant of the *homo economicus* or vice versa. Speaking metaphorically, the *homo economicus* and the *homo socialis* must be distinguished like night and day. This is mainly due to the discontinuous transition in the systemic outcome when the *friendliness* increases beyond a certain critical level (*tipping point*) (as mentioned above). Passing the tipping point is like switching the light on or off, with dramatic impacts on the average payoff (see Fig. 1).

---

90 Arrow and Debreu, "Existence of an Equilibrium for a Competitive Economy;" Debreu, *Theory of Value: An Axiomatic Analysis of Economic Equilibrium*.

91 Coase, "The Problem of Social Cost;" Roca et al., "Coordination and Competitive Innovation Spreading in Social Networks".

92 Dirk Helbing, M. Treiber and N. Saam, "Analytical Investigation of Innovation Dynamics Considering Stochasticity in the Evaluation of Fitness," *Physical Review E* 71.6 (2005): 067101; César A. Hidalgo, B. Klinger, A.-L. Barabasi and R. Hausmann, "The Product Space Conditions the Development of Nations," *Science* 317.5837 (2007), p. 482–487; Scott E. Page, *The Difference: How the Power of Diversity Creates Better Groups, Firms, Schools, and Societies* (Princeton: Princeton University, 2008); Pier-Paolo Saviotti and Koen Frenken "Export Variety and the Economic Performance of Countries," *Journal of Evolutionary Economics* 18.2 (2008), p. 201–218; Philip Cooke, *Complex Adaptive Innovation Systems* (London: Routledge, 2009).

5. SOCIONOMICS: A PROMISING NEW FIELD OF RESEARCH

The existence of two different kinds of people, the *homo economicus* and the *homo socialis*, has relevant practical implications: A second body of work must be developed for the *homo socialis* to supplement the extant economic body of work for the *homo economicus*. I will call the science of the *homo economicus*, *economics*, and the science of the *homo socialis*, *socionomics* or *economics 2.0* in order to reflect that the Web 2.0 is a key factor in promoting a *participatory market society*.[93] It is important to note that our understanding of socionomics does not assume emotionally or irrationally acting agents. In the simulations underlying Figs. 1 and 2, the *homo socialis* decides rationally, trying to optimize the individual utility, but this utility also reflects the externalities of others. In social dilemma situations, the *homo socialis* can out-perform the *homo economicus* through reaching higher payoffs.

Socionomics, and what distinguishes it

In contrast to prevailing work in sociology, socionomics includes the study of suitable institutions and operational principles for future market and exchange systems.[94] Market and exchange systems fitting the *homo economicus* and the *homo socialis* respectively will be different. In the same way as economic market systems are called *economies*, I will call market systems based on socionomic principles *socionomies* or *economies 2.0*. I will also use the term *participatory market societies*, as socionomies bring economic activities together with a social orientation. The participatory character of socionomies will become clearer when I discuss the role of *prosumers*, i.e. of co-producing consumers in Sec. 7.1.

But what might be distinguishing features of socionomies as compared to economies?

- In socionomies, self-regarding optimization is replaced by individual decisions that take into account the implications for others (*externalities*), thereby supporting mutual coordination and cooperation.

- Hence, in socionomies, agents would have cooperative motives in addition to competitive ones. This is sometimes characterized by the term *coopetition*.[95]

---

93 Tim O'Reilly, "What is Web 2.0: Design Patterns and Business Models for the Next Generation of Software," *Communications & Strategies* 1 (2007), p. 17.

94 I would like to point out that socionomics has certain points in common with other emerging research directions such as *evolutionary economics*, *network economics*, *complexity economics*, *behavioral economics*, or *socio-economics*, to mention just a few.

95 See http://en.wikipedia.org/wiki/Coopetition.

– In socionomies, regulation can happen in a bottom-up way, which is called *self-regulation.*

– In socionomies, social dilemma situations can also be overcome without financial incentives and fines, for example, by reputation systems (see Sect. 6.2).

– In socionomies, sharing information and other resources pays off, as this supports coordination and cooperation. As socionomics is still nascent, the above points need to be investigated with simulation models and experiments to identify the power and limitations of the new paradigm of the *homo socialis* and the institutions needed (see Sec. 6).

5.1 *What one can learn from traffic light control*

Like the preferences of economic agents, urban traffic flows are characterized by externalities (as they negatively interfere with each other). Therefore, when discussing different ways of organizing and regulating markets, one can learn much from urban traffic light control. For this purpose, let us treat road intersections as agents that are networked with each other. Then, if we let all intersections optimize the local traffic flows independently of each other, this corresponds to the approach of a *homo economicus.* If the intersections take into account the traffic flows at neighboring intersections in their control decisions, this corresponds to the *homo socialis.* I will also compare both with the approach of a central regulator, who tries to optimize traffic flows in the entire city in a top-down way.

The typical goal of traffic control is to minimize congestion, environmental impact, and travel times. Today, decisions are usually taken top down by a traffic control center, corresponding to a *central regulator.* However, typically, the mobility demand changes more quickly than the control strategy can be adjusted, with the consequence that the optimization of traffic flows often leads to suboptimal results.

In the following paragraphs, I will explain how and why a *self-regulation* approach based on a flexible, adaptive response to local traffic flows can be more successful than the top-down control by a central regulatory authority. To control traffic lights, many cities today use supercomputers trying to identify the optimal solution for the system and to implement it like a *benevolent dictator.* A typical solution creates *green waves* by synchronizing neighboring traffic lights. However, in large cities, even supercomputers are unable to manage strict traffic light optimization in real-time due to the algorithmic complexity of the NP *hard* optimization problem. The number of parameters and the possible combinations of their values are so large that the required computational time *explodes* with the number of traffic lights.

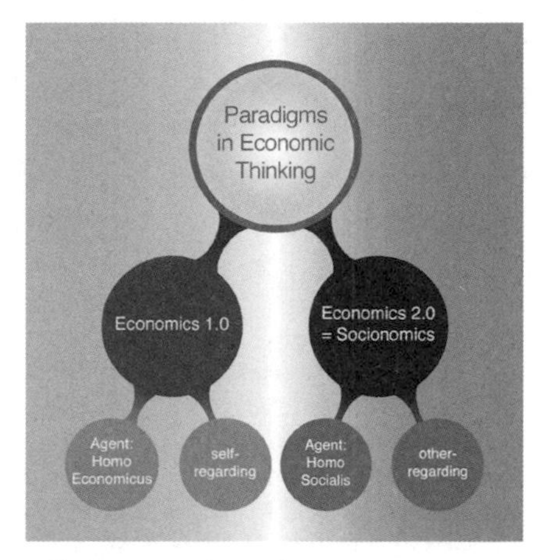

Fig. 3. Illustration of the two distinct economic paradigms.
The difference between the *homo economicus* and the *homo socialis* is that the latter takes into account the interests of others when making decisions, which implies interdependent decisions or *net-worked minds*. The different nature of the *homo socialis* leads to a complex dynamics and another macroscopic outcome than expected for the *homo economicus*.
In the case of public goods problems, for example, interactions of agents with strictly self-regarding preferences will lead to *tragedies of the commons*, while the self-regulation of the *homo socialis* can overcome this undesirable state and foster cooperation, leading to higher individual and social benefits. Due to the different system dynamics and different systemic outcomes, the two types of agents cannot be described by the same body of theory. They require separate sets of literature and different institutions.

Instead, traffic light control schemes are usually optimized offline for representative (average) traffic flows at a certain weekday and time, or for certain events (e.g. football games).[96] The corresponding control schemes are then activated in the situations for which they were optimized. In addition, these may be adapted to the actual traffic situation by extending or shortening green phases. However, at a particular intersection the periodicity (i.e. the order of green phases serving the incoming roads) is usually kept the same during the operation of a control scheme. While the adaptation of traffic phases improves the traffic light control, it must be recognized that the concept of *representative* traffic flows is very misleading: the variability of the number of vehicles waiting at a traffic light is about as large as the average value. This implies a sub-optimal performance of the central regulator approach.

---

96 This is also the approach that the mean value approximation behind the representative agent approach would suggest.

Market System	Centrally Planned Economy	Conventional Market Economy	Participatory Market Society
Agent	Central Planner	Homo Economicus	Homo Socialis
Organization	top down	bottom up	bottom up
Regulation	top down	top down	bottom up

Fig. 4. Comparison of different ways of organizing and regulating markets, based on a (benevolent) *central planner*, the *homo economicus* or the *homo socialis*.
The organization and regulation of the *homo socialis* needs to be distinguished from both, centrally planned an conventional market economies. Hence, I propose to devote a new branch of economics called *socionomics* or *economics 2.0* to the study of the system dynamics, outcomes, and suitable institutions of the *homo socialis*. The related socio-economic systems are *participatory market societies*. These systems are now emerging and spreading due to the Web 2.0, particularly thanks to social media platforms.

Let us now compare the central traffic light control with the *homo economicus* strategy, where each intersection strictly optimizes local traffic flows independently by minimizing average waiting times.[97] We assume that each intersection measures not only the outflows of the incoming road sections, but also receives information about the inflows into these road sections that originate at the neighboring intersections. This information exchange between neighboring intersections allows short-term predictions of the arrival times of vehicles. Hence, the *homo economicus* strategy can respond to this prediction in a way that tries to keep vehicles moving, thereby minimizing waiting times. This self-organizes the local traffic flows, but the principle of the *invisible hand* works only up to a certain traffic volume.[98] Long before the intersection capacity is reached, the average queue length tends to get out of control, as some road sections with small traffic flows are not served frequently enough (see Fig. 7).[99] This will create spillover effects and obstructions of upstream traffic flows, such that congestion quickly spreads over large parts of the city in a cascade-like manner. The resulting state may be compared with a *tragedy of the commons*, as the available intersection capacities are not efficiently used.

---

97 Stefan Lämmer and Dirk Helbing, "Self-Control of Traffic Lights and Vehicle Flows in Urban Road Networks," *Journal of Statistical Mechanics* 2008.04 (2008): P04019.

98 Lämmer and Helbing, "Self-Control of Traffic Lights and Vehicle Flows in Urban Road Networks".

99 In other words: minimum waiting times are generally not compatible with minimum queue lengths and vice versa.

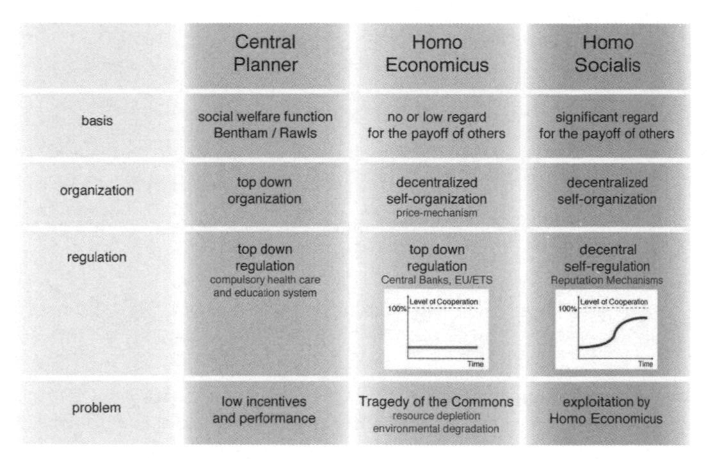

	Central Planner	Homo Economicus	Homo Socialis
basis	social welfare function Bentham / Rawls	no or low regard for the payoff of others	significant regard for the payoff of others
organization	top down organization	decentralized self-organization price-mechanism	decentralized self-organization
regulation	top down regulation compulsory health care and education system	top down regulation Central Banks, EU/ETS	decentral self-regulation Reputation Mechanisms
problem	low incentives and performance	Tragedy of the Commons resource depletion environmental degradation	exploitation by Homo Economicus

Fig. 5. Table highlighting the differences between the nature of a central planner, the *homo economicus*, and the *homo socialis*, including their different system dynamics and systemic outcomes.
The First Theorem of Welfare Economics (Arrow and Debreu, "Existence of an Equilibrium for a Competitive Economy;" Debreu, *Theory of Value: An Axiomatic Analysis of Economic Equilibrium*), which states that competitive markets lead to an efficient allocation of resources, only holds if there is a complete set of markets and all market participants are price-takers, and there are neither externalities nor transaction costs.

As conventional market economies tend to create negative externalities, top-down regulation by a strong state is needed to counter related market failures. Examples of negative externalities are resource depletion and environmental pollution, which eventually lead to the creation of regulatory institutions such as the emission trading scheme or financial regulatory authorities. As regulation tends to be costly for regulators and regulated agents, bottom-up self-regulation is expected to unleash large creative potentials of *networked minds*, which are currently not fully used (see Sec. 7). The performance gains may well be comparable to the gains when moving from a top-down regulation to a conventional market economy building on self-organization (see Sec. 5.1).

Socionomies (also called *economies 2.0* or *participatory market societies*), are characterized by high levels of cooperation, established through self-regulation. To protect a *homo socialis* from being exploited, it is important to know whether one is interacting with another *homo socialis* or a *homo economicus*. Therefore, suitable institutions such as reputation mechanisms need to be established, which allow like-minded agents to interact successfully, and differently-minded agents to cope with each other. Examples for socionomic systems are Wikipedia, Open Streetmap, open source software communities (such as Linux or GitHub), and social media platforms.

Note that the *homo socialis* should not be confused with someone who is redistributing gains. A redistribution strategy (via donations, or taxes and social benefit systems) might be able to *heal* or reduce problems such as an *unhealthy* degree of inequality, but it would usually not be able to avoid or overcome *tragedies of the commons*. Hence, redistribution is inefficient and often not a suitable solution for social dilemma situations. The main difference to the *homo economicus* is that the *homo socialis* decides more *responsibly* or *wisely*, considering not only his own interests, but also those of others when taking self-determined decisions. If enough people do this, other-regarding preferences pays off.

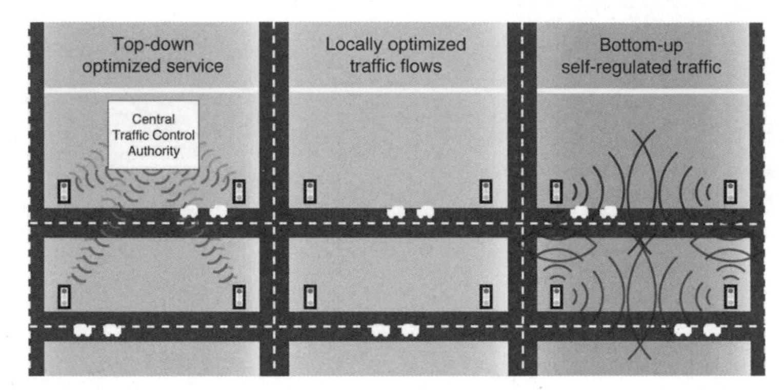

Fig. 6. Schematic illustration of three kinds of traffic light control: a *central regulator*, a strategy minimizing local waiting times (*homo economicus* approach) and a strategy considering externalities on neighboring intersections (*homo socials* approach).

In order to outperform the central regulator over the whole range of traffic demands (utilizations) that an intersection can handle, one must extend the *homo economicus*-based self-organization to a self-regulatory approach. Specifically, the rule of waiting time minimization must be combined with a second rule, according to which a vehicle queue must be cleared immediately, whenever it reaches a critical length (a certain percentage of the road section). This second rule avoids spillover effects that would obstruct neighboring intersections and therefore acts in an other-regarding way, as a *homo socialis* would do. The self-regulation strategy uses occurring gaps as opportunities to serve other traffic flows. So, rather than planning green waves by synchronizing traffic lights in a certain rhythm, self-regulated traffic lights take into account the traffic situation at neighboring intersections, just as a *homo socialis* would do. This enables the coordination of neighboring traffic lights which can spread over many intersections.

In summary, the self-regulation approach coordinates traffic lights and traffic flows based on local interactions. While the self-regulation is based on a decentralized bottom-up approach rather than a centralized top-down approach, it surprisingly reaches a superior performance.[100] Studies comparing self-regulated traffic lights with state-of-the-art traffic control approaches show that shorter waiting times can be simultaneously reached for individual traffic, public transport, pedestrians, and bikes.[101] Self-

—

100 Ibid.
101 Ibid.

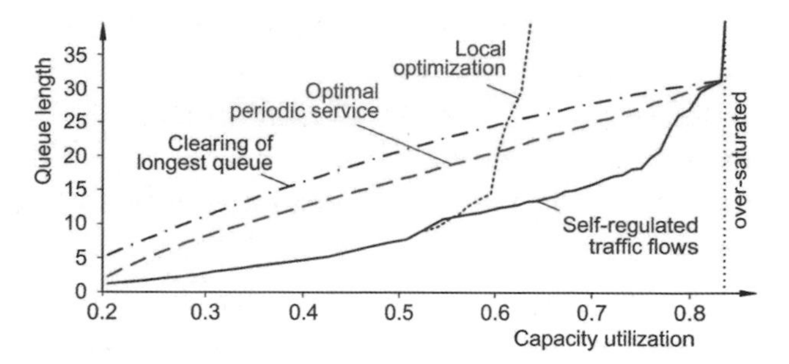

Fig. 7. Comparison of the performance of a central regulator, local optimization of traffic flows as a *homo economicus* would do, and other-regarding self-regulation, considering the impact on neighboring intersections.
The graphics shows average queue lengths resulting for these different traffic control approaches (after Stefan Lämmer and Dirk Helbing, "Self-Stabilizing Decentralized Signal Control of Realistic, Saturated Network Traffic," *Santa Fe Institute Technical Report*, 10-09-019 (2010). Always clearing the longest queues performs worse than a periodic service that is optimized in a top-down way, as a central regulator would do. A local minimization of waiting times, corresponding to a situation where each intersection acts like a *homo economicus*, outperforms a periodic service at low utilizations of the theoretical intersection capacity, but it creates long queues and spill-over effects at high utilizations. A decentralized self-regulation with other-regarding decision rules, corresponding to the approach of a *homo socialis*, performs best. It also minimizes waiting times locally, but this principle is over-ruled by the clearing of critical queues, which might block neighboring intersections by spill-over effects. The self-regulation approach also outperforms an optimized cyclical service, where the green times are adapted to the respective traffic situation.

controlled traffic also reduces environmental pollution without having to impose travel restrictions, and it also makes travel times more predictable.

From the above example, we can draw a number of important conclusions:

1. In complex systems with strongly variable and largely unpredictable dynamics, bottom-up self-regulation can outperform top-down optimization by a central regulator (an urban traffic control center). This also holds when input data are comprehensive and reliable.

2. Strict local optimization according to the principles of a *homo economicus* may create a highly-performing system within a certain parameter range, but tends to fail when interaction effects by traffic flows of neighboring intersections are strong and the optimization at each intersection is performed separately, in a self-regarding way.

3. When taking into account the situation of interaction partners (*neighbors*), as an other-regarding *homo socialis* would do, a high system performance can be reached even for strong interaction effects because of coordination between neighbors.

In conclusion, a central regulator fails to be efficient for lack of computational capacity, which is required to handle the algorithmic complexity of a NP-hard optimization problem within large systems in real-time. Applying the principle of the *homo economicus* on a local level fails due to lack of coordination. In contrast, the approach of the other-regarding *homo socialis* can overcome both problems by combining self-organization with self-regulation. A self-regulation based on suitable local interactions can produce resource-efficient solutions delivering high system performance.[102] These solutions also tend to be resilient to perturbations (such as accidents, road works, etc.). Again, the *homo socialis* can make the principle of the *invisible hand* work under conditions, where it fails for the independent optimization attempts of a *homo economicus* (see Fig. 7).

Many of the above conclusions also seem to be relevant for socio-economic systems, when agents have incompatible interests that cannot be satisfied at the same time.

6. INSTITUTIONS FOR THE *HOMO SOCIALIS*

6.1 Mechanisms to overcome social dilemma situations

Institutions for market systems should not just work under ideal (*good weather*) conditions, but also under difficult ones. They should be suited not only for best-case scenarios, but also for worst-case scenarios, in as much as this is possible. Therefore, in the spirit of the self-regulation approach, we are looking for:

1. rules to handle externalities of decisions, protecting others from exploitation and compensating them for damage,

2. frameworks to coordinate between individuals and projects with different preferences and interests.

In particular, we are searching for institutions that work in the case of social dilemma situations. From game theory, it is known that cooperation is endangered by interaction with defectors.[103]

As the *homo economicus* is always expected to defect when deciding according to a best response rule (maximizing the payoff given the current behaviors of the neighbors),

---

102 The recent trend towards replacing many signalized intersections by roundabouts and changing urban spaces regulated by many traffic signs in favor of designs supporting a considerate interaction and self-organization of different traffic participants suggests an on-going paradigm shift from regulation to self-regulation (Ben Hamilton-Baillie, "Shared Space: Reconciling People, Places and Traffic," *Built Environment* 34.2 (2008), p. 161–181.

103 Ernst Fehr and Simon Gachter, "Altruistic Punishment in Humans," *Nature* 415.6868 (2002), p. 137–140.

cooperation by the *homo socialis* must be somehow protected from exploitation by the *homo economicus*. In other words, the *homo socialis* is not expected to thrive in institutional settings designed for the *homo economicus* such as homogeneous global markets, which destabilize cooperation in social dilemma situations.[104] I would like to stress again that any situation, in which a market participant can be cheated, establishes a social dilemma situation. Nevertheless, there are solutions to the problem.

Evolutionary game theory has identified a number of biological and social mechanisms that can create suitable institutional settings for cooperation in social dilemma situations. These include genetic favoritism and direct reciprocity ("I help you, if you help me"),[105] but also punitive institutions (such as the criminal justice system, and legal and regulatory authorities). However, all these mechanisms have undesired side effects. Genetic favoritism impedes an open global exchange and leads to ethnic conflict. Direct reciprocity is hardly compatible with global openness to new trading partners, and could also incentivize corruption. Punishment institutions are costly and create inefficiencies, as discussed in Sec. 4. In Sec. 3, cooperation spreads due to a naturally resulting spatial segregation between cooperators and defectors, when intergenerational migration is low.[106] Another mechanism would be based on reputation systems,[107] which transfers the functional principle of social communities to a globalized world. Such reputation systems work also in a decentralized way.[108]

—

104 Dirk Helbing, W. Yu and H. Rauhut, "Self-Organization and Emergence in Social Systems: Modeling the Coevolution of Social Environments and Cooperative Behavior," *The Journal of Mathematical Sociology* 35.1–3 (2011), p. 177–208; Dirk Helbing, "Globally Networked Risks and How to Respond".

105 Nowak, *Evolutionary Dynamics*.

106 Grund et al., "How Natural Selection Can Create Both Self-and Other-Regarding Preferences, and Networked Minds".

107 Rosaria Conte and Mario Paolucci, *Reputation in Artificial Societies. Social Beliefs for Social Order* (New York: Springer, 2002); Manfred Milinski, Dirk Semmann and Hans-Jürgen Krambeck, "Reputation Helps Solve the Tragedy of the Commons," *Nature* 415.6870 (2002), p. 424–426.

108 Sep Kamvar, M. T. Schlosser and H. Garcia-Molina, "The Eigentrust Algorithm for Reputation Management in p2p Networks," *Proceedings of the 12th International Conference on World Wide Web, ACM* (2003), p. 640–651; Sonja Buchegger and Jean-Yves Le Boudec, "A Robust Reputation System for Peer-to-Peer and Mobile Ad-Hoc Networks," *Proceedings of P2PEcon* (2004); Li Xiong and Ling Liu, "Peertrust: Supporting Reputation-Based Trust for Peer-to-Peer Electronic Communities," *IEEE Transactions on Knowledge and Data Engineering* 16.7 (2004), p. 843–857; Runfang Zhou and Kai Hwang, "PowerTrust: A Robust and Scalable Reputation System for Trusted Peer-to-Peer Computing," *Parallel and Distributed Systems, IEEE Transactions on* 18.4 (2007), p. 460–473; Jochen Mundinger and Jean-Yves Le Boudec, "Analysis of a Reputation System for Mobile Ad-Hoc Networks with Liars," *Performance Evaluation* 65.3–4 (2008), p. 212–226.

6.2 User-centric, multi-criteria reputation systems and incentives

The purpose of the reputation system is to avoid costly transaction failures arising from mismatched expectations regarding the kind of economic exchange.[109] In other words, reputation systems can help different types of agents such as the *homo socialis* and the *homo economicus* to cope with each other. Reputation systems are currently spreading on the Web. A typical example is eBay. Reputation and recommender systems are now used to evaluate goods, service providers and news. It has been shown that eBay sellers with a higher reputation can sell products at a higher price, i.e. a good reputation pays off.[110] This incentivizes reputation-enhancing behavior. In addition to price- based competition, recommender and reputation systems also endorse quality-based competition, i.e. features such as responsible environmental and social production conditions become more relevant and rewarding.

An important question to ask is: who would decide the reputation of someone or something? To successfully cope with a system of high complexity (see Sec. 5.1), it seems necessary to build on a bottom-up self-regulation approach. Therefore, the reputation should be decided decentrally, by each decision-maker, based on his or her own preferences and values. In other words, it should not be a company or another institution that determines the reputation, but the individuals.[111] We should also consider that criteria determining high quality for some may represent poor quality for others. Thus ideally, recommender systems should be multi-dimensional, based on a variety of criteria. Decision-makers can then define filters determining reputation based on their own criteria.[112] Moreover, people could share their own reputation filters with others, who may

---

109 Fabian Winter, Heiko Rauhut and Dirk Helbing, "How Norms Can Generate Conflict: An Experiment on the Failure of Cooperative Micro-motives on the Macro-level," *Social Forces* 90.3 (2012), p. 919–946.

110 Wojtek Przepiorka, "Buyers Pay for and Sellers Invest in a Good Reputation: More Evidence from eBay," *Journal of Socio-Economics* 42 (2013), p. 31–42.

111 The problems of recommender systems, which cannot be configured by the users themselves, have been discussed in the book *Filter Bubble* (Eli Pariser, *Filter Bubble* (München: Carl Hanser, 2012)). One of them is the manipulation of decision-makers, because the *wisdom of crowds* can be seriously undermined by feeding back opinions of others (Jan Lorenz, H. Rauhut, F. Schweitzer and D. Helbing, "How Social Influence can Undermine the Wisdom of Crowd Effect," *Proceedings of the National Academy of Sciences of the United States of America* 108.22 (2011), p. 9020–9025). This can reduce socio-diversity as well (Dirk Helbing, "Google as God? Opportunities and Risks of the Information Age," (2013) *Essay available at* http://arxiv.org/abs/1304.3271) and, thereby, affect the functionality of fundamental institutions that societies are built on. Democratic and market institutions require independent decision-making to function well.

112 As a consequence, someone may have a good reputation in some areas and no or a bad reputation in others, depending on the values and viewpoints of the potential interaction partners. Therefore, a user-centric, multi-criteria reputation system can help one to find like-minded people. Note that reputation data do not have to be public, but may be provided on request.

modify them according to their own perspective. Therefore, opening up recommender data to everyone would enable the evolution of an *information ecosystem*, in which customized information filters would steadily improve over time. A similar reputation principle could also be applied to money.

6.3 Qualified money

One interesting – and somewhat speculative – question is what would happen, if we applied the principles of reputation systems to money itself,[113] i.e. if each unit of money could earn a reputation, depending on its origin and transaction history.[114] Then, units of money could be treated as separate stocks. Thus money would be related not just with a quantity, but also with some qualities. This would make money multi-dimensional, akin to feedback and exchange systems in biological and ecological systems, or also social systems.[115] I call this concept *qualified money*.

Note that the one-dimensionality of today's money has a number of problematic features. It basically implies that there can only be ups and downs. The bubbles and crashes that financial systems experience over the last hundreds of years[116] may be a consequence of this. For a complex dynamic system such as the financial system to work well, it is important that there are enough parameters to allow the system to adapt.[117] Currently, the financial system suffers from a lack of such parameters.

If we had *qualified money*, a conversion factor would apply, which would determine the value of money together with its quantity. The conversion factor would depend on the qualities of the respective money units, which would be given by multiple reputation values. Hence, the conversion factors establish adaptive parameters. Therefore,

---

113 Tim Moreton and Andrew Twigg, "Trading in Trust, Tokens, and Stamps," Proceedings of the "First Workshop on Economics of Peer-to-Peer Systems," http://www.cs.ox.ac.uk/people/andy.twigg/pubs/2003-trading-trust.pdf (accessed September 2013).

114 I would like to underline that this section has currently a preliminary and speculative character, as it first needs to be grounded on computational and empirical evidence before it can be practically applied. However, the exploration of complementary, alternative, and backup financial systems seems to be in place.

115 Alan Fiske, *Structures of Social Life: The Four Elementary Forms of Human Relations: Communal Sharing, Authority Ranking, Equality Matching, Market Pricing* (New York: Free Press, 1993).

116 Glyn Davies, *A History of Money: From Ancient Times to the Present Day* (Cardiff: University of Wales Press, 2002); David Graeber, *Debt: The First 5,000 Years* (London: Melville House, 2011).

117 Note that trying to reach many different, non-aligned goals with just one control variable is an *ill-defined* (unfeasible) control problem. Assume that one tries to encourage k desirable behaviors and discourage l undesirable ones with corresponding positive and negative tax incentives, then the overall effect will be rather unspecific and most of the goals are unlikely to be reached.

Euros in a certain country could gain a higher or lower reputation (and value) than in other countries. If a country were to suffer from an economic depression, the conversion factor would decrease, and the corresponding devaluation of money would help the country to solve its problems by inflation. In other words, the international financial system would have enough degrees of freedom for self-regulation to work.

As indicated before, the qualifiers could also be made dependent on the transaction history. If money were history-dependent, then money generated in certain ways (e.g. by environmental-friendly production) would gain additional value, if a customer cares about it. This would create incentives to invest in quality, not just quantity. Therefore, the downward spiral leading to *tragedies of the commons* could be overcome. There would be a competition for money of higher quality. Note that the properties entering the qualifiers (and those that are not made transparent or evaluated) would be decided in a bottom-up way, first of all by the confidentiality settings of the owner and second by letting potential customers choose their own quality filters, as described for the multidimensional reputation system before. This would establish an implicit negotiation process between sellers and buyers, i.e. it would give customers more influence on price formation than just by a decision to buy a product or abstain from it. In this way, customers would become a *voice* allowing them to inform producers about the qualities they value and the factors they care less about. If this would influence the corresponding conversion factors strongly enough, it would be an effective way of improving quality standards. In such a way, *qualified money* could also help to increase sustainability.

6.4 *Platforms for participation, exchange and mutual compensation*

Another important institution for the *homo socialis* are participatory platforms that support a trusted and fair exchange,[118] as required to coordinate with each other and to consider the interests and preferences of others. Such platforms should provide tools for cooperation, allowing people to set up joint projects, communicate and collaborate. Ideally, the platform would support everything from the scheduling of processes to supply chain management and accounting. The functionality should be easy to use at an affordable price with open and interoperable interfaces, while providing secure information storage and exchange for sensitive data. New information and communication technology (ICT) and the data they produce and manage will also make it possible to determine

---

118 Yves-Alexandre de Montjoye, Y.-A. de Montjoye, S. S. Wang, A. S. Pentland, D. T. T. Anh, A. Datta, K. W. Hamlen, L. Kagal, M. Kantarcioglu, V. Khadilkar, K. Y. Oktay, et al., "On the Trusted Use of Large-Scale Personal Data," *Data Engineering* (2012), p. 5.

and quantify harmful interactions, and to reduce them by compensation mechanisms (such as appropriate payments). This possibility arises, in particular, as the *Internet of Things* spreads.

The FuturICT initiative has recently proposed to build a "Global Participatory Platform," which would eventually allow everyone to set up collaboration projects with others, underpinned by Open Data.[119] Such a platform could also be used for eGovernance applications such as the payment of taxes, health and other insurances. From the perspective of self-regulation, it seems to make sense that the people involved in participatory decision-making would be those who are significantly affected by the respective externalities. In the future, this principle might be applied to economic and political decision-making alike.

6.5 Innovation accelerator

To promote innovation, it is important to have incentives that encourage investments in innovation. Today, this is done by protecting intellectual property by patents. It often appears, however, that patents can also obstruct innovation. It would therefore be desirable to find a mechanism to reward innovations automatically, whenever someone's idea is used.[120] If we had a way to measure the value of micro-innovations, we could innovate in a much more collaborative and open way. Whoever has a good idea could contribute to a difficult invention and get a reward for it. Building on crowd sourcing and swarm intelligence in such a way could largely speed up innovation, and also lead to higher-quality results.[121] Rewards could either be micropayments per use (as *Spotify* does it, for example) or non-monetary incentives (such as citations in science or reputation points), or both.

—

119 Simon Buckingham Shum, K. Aberer, A. Schmidt, S. Bishop, P. Lukowicz, S. Anderson, Y. Charalabidis, J. Domingue, S. Freitas, I. Dunwell, B. Edmonds, F. Grey, M. Haklay, M. Jelasity, A. Karpištšenko, J. Kohlhammer, J. Lewis, J. Pitt, R. Sumner and D. Helbing, "Towards a Global Participatory Platform: Democratising Open Data, Complexity Science and Collective Intelligence," *EPJ Special Topics* 214.1 (2012), p. 109–152.

120 In this connection, Altmetrics (http://www.altmetric.com/) measuring page views and other web indicators might play an important role in the future.

121 Frank van Harmelen, G. Kampis, K. Borner, P. van den Besselaar, E. Schultes, C. Goble, P. Groth, B. Mons, S. Anderson, S. Decker, C. Hayes, T. Buecheler and D. Helbing, "Theoretical and Technological Building Blocks for an Innovation Accelerator," *EPJST* 214.1 (2012), p. 183–214.

6.6 Socionomics, or Economics 2.0

Last but not least, one important institution to promote the *homo socialis* would be a scientific discipline to study the implications of this concept and suitable institutional settings.[122] As pointed out before, this calls for a field such as *socionomics* or *economics 2.0*.[123] Note that one particular challenge will be to study the principles of successful self-regulation within complex systems. This will have to go beyond mechanism design,[124] taking into account evolutionary principles[125] and thereby endogenizing innovation into the system dynamics. Inspiration could also be drawn from ecological,[126] immune,[127] or social systems.[128]

Evolutionary principles would be able to identify improved rules for self-regulation in a truly bottom-up way. The case of traffic light control (see Sec. 5.1) has shown that small changes of rules may fundamentally change the systemic outcome and performance (such as large-scale congestion as compared to free traffic flows). When a system is sensitive to details, self-regulation must include rules that stabilize the application of successful rules (e.g. avoid a self-regarding modification of rules). However, over-rigidity would affect innovation and, therefore, the ability to find even better sets of rules. Evolutionary systems are therefore characterized by checking out diverse variants (thanks to innovation-generating *mutations*) in combination with autocatalytic or reproduction mechanisms ensuring that better variants spread more quickly. As a consequence, a decentralized approach is an important precondition for success, while the implementation of homogeneous rules in large areas is expected to reduce diversity and innovation.

---

122 Andrew Schotter, *The Economic Theory of Social Institutions* (Cambridge: Cambridge University Press, 1981); Christian Schubert and G. von Wangenheim, *Evolution and Design of Institutions* (London: Routledge, 2006).

123 Norbert Häring and Olaf Storbeck, *Economics 2.0* (Basingstoke: Palgrave Macmillan, 2012).

124 Leonid Hurwicz and Stanley Reiter, *Designing Economic Mechanisms* (Cambridge: Cambridge University Press, 2006).

125 David E. Goldberg and Kumara Sastry, *Genetic Algorithms: The Design of Innovation* (New York: Springer, 2010).

126 Haldane and May, "Systemic Risk in Banking Ecosystems".

127 Dario Floreano and Claudio Mattiussi, *Bio-inspired Artificial Intelligence* (Boston, Massachusetts: MIT Press, 2008).

128 Helbing, *Social Self-Organization.*

7. ECONOMIES 2.0 AS PARTICIPATORY MARKET SOCIETIES

In the previous sections, I have discussed some theoretical arguments supporting a new organization of market exchange. In this section, we will see that markets have already started to re-organize in the suggested direction. For example, reputation and recommender systems are now found all over the Web 2.0. Likewise, social media are providing participatory platforms, which are increasingly used to set up collaborative projects. The example of Facebook demonstrates that even with free platforms one can generate considerable economic value. The next paragraphs give a better idea of how future information and communication technologies are expected to change the lives of workers, consumers and entrepreneurs.

7.1 Prosumers and the future role of entrepreneurs

Today, we are still living in a world of slowly adapting institutions which (in the best case) are trying to take optimal decisions for many people, based on representative data, for example political parties or companies. Although they have bottom-up elements, a great deal of decision-making is done in a top-down manner. However, as systems become more complex, they will require more bottom-up elements or they will otherwise perform poorly (see Sec. 5.1) or even destabilize over time.[129]

Digital technologies are now enabling more flexible ways of organization. In particular, people may use social media platforms to organize *projects* in a bottom-up way. In principle everyone could do this, given the required technical and social skill sets. This development could eventually turn consumers into *prosumers*,[130] i.e. consumers who are co-creating products they buy (and sell).[131] 3D printer technology, for example, will enable local production by small teams or individuals who may sell their products to friends, colleagues or the rest of the world. Rather than just specifying the color and individual features of a car when ordering it, we may soon design some of its components and commission their production. One may even set up a team of designers, engineers, marketing people, and other specialists to design an own car with components produced by other companies or with new components commissioned from

---

129 Helbing, "Globally Networked Risks and How to Respond".

130 See http://en.wikipedia.org/wiki/Prosumer and http://www.ted.com/talks/alastair parvin architecture for the people by the people.html.

131 Bill Quain, *Pro-Sumer Power II! How to Create Wealth by Being Smarter, Not Cheaper, and Referring Others to Do the Same*, International Network Training Institute, Inc (2008).

home. That is, old-style factories and socionomies based on collaborative projects are expected to work hand in hand. To a certain extent, *projects* of the above kind are already in existence today, for example open source software projects. Many such projects are driven by volunteers or employees of companies, who rely on open source components and want to get their required features implemented. In these projects, the development takes place bottom-up and open. The *open source ecosystem* is based on a number of ingredients such as *viral* open source licenses, which ensure that those using open source code in their own software will also have to make it available to others. Therefore, software licenses (such as the GNU General Public License) reward other-regarding behavior as it characterizes the *homo socialis*. In the context of open source development, the GitHub platform[132] has recently become very popular among software developers. The platform indicates who has contributed how much, thereby creating incentives for contributing. Thus, everyone can benefit from a growing set of open source software.

The digital economy will open up infinite opportunities for new products. We might even say that the information age enables infinite dimensions of creativity. This has very interesting implications for the resulting market structures. Today, we have a few core businesses and some peripheral business activities. In the future, however, one may expect that projects and peripheral products will dominate market activities.[133] This means that many more products will be individually customized. Note also, that projects imply new forms of work and employment.[134] But if suitable institutions exist, projects could create a large number of jobs in the future. Therefore, socionomies could not only unleash an age of creativity, but also provide novel ways to overcome the current unemployment misery.

It is likely that, over time, companies, political parties, and other established institutions will be complemented by *projects* as more flexible forms of organization. Then, the future role of entrepreneurs will be to set up and coordinate such projects, and organize the necessary support. Once completed, a project would terminate, and the previously involved people would look for new projects to coordinate or participate in. In this way,

---

132 https://github.com.

133 One may imagine this as the core of a sphere as compared to its surface area. Note, however, that the relationship between the surface area A of an n-dimensional sphere and its volume V is $V=rA/n$, see http://en.wikipedia.org/wiki/Sphere. That is, the higher the dimensionality of a market, the more happens in the peripheral (surface) area.

134 Amazon Mechanical Turk, for example, is a platform that matches tasks and workforce.

a *participatory market society* would emerge.[135] Everyone could be both a coordinator and a participant of several projects, which provides opportunities to influence issues one cares about. The fluid, project-based organization of socio-economic activities might also be a good solution to the Peter Principle,[136] according to which people get promoted until they end up in a position which overstrains their abilities.

Projects will also imply more self-determined and exciting work. To unleash creativity, decision-making needs to take place differently from the way it works today. Currently, many decisions are majority-based or top-down decisions. In the first instance, a majority decision is a compromise, which is often disappointing to all (the *common denominator*). In the second case, someone imposes decisions on others, which are potentially against their preferences. This creates advantages for some and disadvantages for others. However, it should be remembered that diversity is the main driving force of innovation in evolutionary systems.[137] It would, therefore, be better to allow projects to find their own rules.[138] People could then easily find projects and socio-economic environments that fit their individual preferences, interests and needs.

8. CONCLUSIONS AND OUTLOOK

Globalization and technological revolutions have created new levels of interdependencies, interconnectedness and complexities, which have the potential to destabilize our techno-socio-economic-environmental system(s) on a global scale.[139] It is therefore critical that new approaches are found to stabilize global networks and counter systemic instability.

The current trend seems to point towards *surveillance societies* with strong punitive elements (*punitive societies*). However, such a top-down approach would endanger pri-

---

135 Note that *cooperatives*, which are a common form of business organization in Switzerland and elsewhere, fit a participatory market society well, but *projects* are expected to be more short-lived and flexible.

136 Laurence Peter and Raymond Hull, *The Peter Principle* (New York: HarperCollins Publishers, 1969).

137 Helbing et al., "Analytical Investigation;" Hidalgo et al., "The Product Space Conditions the Development of Nations;" Page, *The Difference*; Saviotti and Frenken, "Export Variety and the Economic Performance of Countries;" Cooke, *Complex Adaptive Innovation Systems*.

138 Future ICT systems will support and simplify coordination between people with different interests and backgrounds. *Inter-cultural adapters* will help to overcome the need for everyone to agree on the same rules and principles. Reputation systems will promote a proper quality (see Sec. 6.2).

139 Helbing, "Globally Networked Risks and How to Respond".

vacy, socio-diversity and innovation. It also implies major risks that personal data will sooner or later be misused and that a transition to a totalitarian state could occur.[140] Furthermore, recent findings suggest that crime cannot be eliminated by large fines,[141] contrary to what classical rational choice models suggest, and there is little evidence that surveillance by CCTV cameras can reduce crime in a systematic and statistically significant way.[142, 143] Reputation-based interactions could be a more sustainable basis of a stable society. In fact, another recent trend points towards *reputation societies*, but the underlying reputation systems should be user-centric, based on multiple criteria, and run in a decentralized way.[144]

I expect that decentralized self-organization and self-regulation approaches can deliver solutions for complex dynamic systems that are far superior to our existing way of designing and operating systems. However, self-regulation does not mean that one can choose the rules one likes. In fact, as could be seen for in urban traffic light control example, self-regulation on the component level generally does not work, but it must be other-regarding (which requires a certain amount of trustable information exchange).[145]

To be successful, societies must be able to resolve *social dilemma situations*, in particular free-riding must be efficiently contained. In social dilemma situations, cooperative behavior would be favorable for everyone, but exploiting others creates even higher payoffs. Such situations often lead to *tragedies of the commons*. Environmental pollution, overfishing, global warming, free-riding, tax evasion, and the exploitation of our social benefit systems are typical results. Problems like these are expected in particular for the *homo economicus*, who takes optimally self-regarding decisions. In spite of this, the *homo*

---

140 Helbing, "Google as God? Opportunities and Risks of the Information Age".

141 Matjaz Perc, M., Karsten Donnay and Dirk Helbing, "Understanding the Recurrent Nature of Crime as an Evolutionary, Systemic Phenomenon," PLOS ONE 8.10 (2013): e7606.

142 Martin Gill and Angela Spriggs, *Assessing the Impact of CCTV*, London Home Office Research, Development and Statistics Directorate (2005.)

143 UK has surprisingly high crime rates in Europe, given the high coverage with CCTV cameras. A recent terror attack in London intentionally committed below a CCTV camera also questions whether surveillance can be an efficient strategy to stabilize society.

144 Zhou and Hwang, "PowerTrust," Mundinger and Le Boudec, "Analysis of a Reputation System".

145 Also note that the most efficient way of self-regulation may contain hierarchical elements, combining top-down and bottom-up elements in a favorable way. So, at least in the next decades, I do not expect reputation systems to replace top-down elements of social organization, but rather to complement them. Also, *qualified money* would exist next to our current kind of money. However, I expect the relative proportion of self-regulation (bottom-up organization) to grow substantially over time.

economicus is still a foundation of mainstream economic thinking and many policies today.

As non-linear interaction and network effects become more and more important for today's decision-making,[146] a new approach is needed. The approach of an other-regarding *homo socialis* now seems to be a more promising theoretical starting point than the independently deciding *homo economicus*. Like the *homo economicus*, the *homo socialis* results from the merciless forces of natural selection, but it can overcome *tragedies of the commons* and, therefore, achieve higher payoffs.

To understand this, it is important to study economic systems from the viewpoint of complexity theory. We have seen that the combination of non-linear interactions with randomness (such as *mutations*) can produce counter-intuitive emergent phenomena. Complexity theory could also provide a new perspective on socio-economic systems generally.[147] The policy implications can be quite different from those for the *homo economicus*. Therefore, one can expect novel insights and fundamentally new solutions to old problems, including (over-)regulation, innovation, financial crises, unemployment, and sustainability.

As Albert Einstein said: "We cannot solve our problems with the same kind of thinking that created them." In the past, there has been a steady struggle between economic activities and social motives, and between bottom-up market organization and political top-down regulation. Such regulation serves to avoid exploitation, *tragedies of the commons*, and market failures, whenever the self-organization of the *homo economicus* does not lead to desirable outcomes. The *homo socialis* bridges the gap between these apparently incompatible sides. However, as the interdependent decisions of other-regarding, *networked minds* cause a complex system dynamics and different outcomes on the macro-level, a new scientific discipline is needed, called *economics 2.0* or *socionomics*. This research field will have to study the implications of the *homo socialis* and supportive institutional settings, in particular successful principles of self-regulation.

The new conceptual framework may be characterized as follows:

1. The *homo socialis* takes self-determined, but responsible decisions, caring about their impact on others. While this fits well with Kant's moral imperative[148] and with the

---

146 Frank Schweitzer, Giorgio Fagiolo, Didier Sornette, Fernando Vega-Redondo, Alessandro Vespignani and Douglas R. White, "Economic Networks: The New Challenges," *Science* 325.5939 (2009), p. 422–425; Paul Ormerod, *Positive Linking: How Networks Can Revolutionise the World* (London: Faber and Faber, 2012).

147 Helbing and Kirman, "Rethinking Economics Using Complexity Theory".

148 See also the somewhat related work of John E. Roemer, "Kantian Equilibrium," *Scandinavian Journal of Economics* 112.1 (2010), p. 1–24..

values promoted by many religions, it does not need to be based on ethical or religious grounds.[149] The concept is built on scientific insights showing that other-regarding behavior helps to coordinate between incompatible interests, leading to better individual outcomes over a long-enough time horizon.

2. As the *homo socialis* is able to get more payoff under conditions that are highly competitive, the concept of participatory market societies is not based on an idealistic approach. There is also much empirical evidence in our digital economy for this.

3. The approach discussed in this paper promotes social order and social welfare, but also sustainable prosperity, individual freedom, diversity and pluralism. It is very different from the concept of a central regulator discussed in Sect. 5.1. In particular, it does not impose a solution, which it considered best for everyone, on others, be it equality or something else.

4. The new approach builds, rather, on suitable principles of self-organization and self-regulation based on decentralized decisions. This will allow one to cut down today's overregulation, thereby creating larger individual freedoms.

5. Efficient ICT-enabled reputation systems and *qualified money* might support such self-regulation, if they are open, user-centric, participatory, and based on sufficiently many criteria. They could create an institutional setting allowing the *homo socialis* and the *homo economicus* to coexist.[150] *Qualified money* would transfer established principles of value exchange from stock markets to units of money. In particular, it would consider that information is a precondition for markets to function, while today's money is memory-less, thereby cutting away important information that may allow financial systems to be more adaptive and resilient. I expect the best solution to be a market system in which conventional and qualified money would exist in parallel with suitable conversion rules.[151]

—

149 Robert Nelson, *Economics as Religion: From Samuelson to Chicago and beyond* (Pennsylvania: Penn State Press, 2001).

150 Note that both the *homo economicus* and the *homo socialis* are needed for a diverse and pluralistic society to function well. For example, the *wisdom of crowds* should work well for independent decision-makers (the *homo economicus*), while it might be undermined by social influence (as is characteristic for the *homo socialis*) (Lorenz et al., "How Social Influence can undermine the wisdom of crowd effect"). However, the *homo socialis* seems to be better in dealing with social dilemma situations. So, both kinds of behavior have their advantages and disadvantages, depending on the respective context.

151 The reputation system will be described in more detail elsewhere. It may allow users to determine themselves whether they want to share the reputation information about them with others and with whom. It should also establish a suitable balance between traceable and anonymous (information or financial) exchange, thereby limiting misuse of anonymity and stimulating responsible action on the

Web 2.0 is a major driver of the transition from the conventional market economy (*economy 1.0*) to a *participatory market society* (*economy 2.0*). The change is largely fueled by social media platforms, 3D printers enabling local production, and *Big Data* celebrated as the *Oil of the 21st Century*. It is also driven by an increasing level of complexity, which can only be mastered by a higher level of decentralization, as implied by the concept of self-regulation.[152]

This trend towards more decentralization is already visible in the way the Internet is organized, the way smart grids are now being run, and the way modern traffic systems will be managed. For socio-economic systems, the trend towards decentralization implies the participation of more people in social, economic and political affairs. A historical perspective also suggests that more and more people gain influence on decision-making processes as centuries progress.[153]

In fact, the digital revolution is already reshaping our economy.[154] Social media have become highly successful and influential. App Stores have created an ecosystem of millions of apps. Even when provided for free, they can create considerable economic value. Furthermore, the *sharing economy* is currently generating growth rates of 20%. So, participatory market societies have great potential.

Humanity has now the opportunity to enter an era of creativity and prosperity built on social principles and enabled by modern ICT systems. It just takes a proper understanding of socionomic systems to determine and establish suitable institutions such as open, user-centric, multi-criteria reputation systems. In the past, humans have created institutions such as markets, roads, courts, museums, schools, libraries, and universities to benefit our society and economy. It is now time to create the right institutions

—

one hand, while providing individual freedoms on the other hand. Finally, the reputation system could include the feature of *forgetting* and *forgiving*.

152 Andrew Haldane at the Bank of England puts it like this: "Modern finance is complex, perhaps too complex. Regulation of modern finance is complex, almost certainly too complex. That configuration spells trouble. As you do not fight fire with fire, you do not fight complexity with complexity. Because complexity generates uncertainty, not risk, it requires a regulatory response grounded in simplicity, not complexity" (Andrew G. Haldane and Vasileios Madouros, *The Dog and the Frisbee*, Speech presented at the Federal Reserve Bank of Kansas City's Jackson Hole economic policy symposium (2012)).

153 This may be seen as a consequence of time scale separation, which is a precondition for power in hierarchical systems (Helbing, Social Self-Organization; Helbing, "Globally Networked Risks and How to Respond"). One may imagine the result of this process to be a *Swiss basic democracy plus*, enabled by future ICT systems.

154 Andreas Pyka and Horst Hanusch, *Applied Evolutionary Economics And The Knowledge-Based Economy* (Cheltenham: Edward Elgar Publishing, 2006); Loet Leydesdorff, *The Knowledge-Based Economy* (Boca Raton:Universal Publishers, 2006); Häring and Storbeck, *Economics 2.0*.

for the information society, supporting socio-economic life in the 21st century. If the right decisions are made, these institutions will unleash the creativity of our *networked minds,* thereby opening the door to an age of sustainable prosperity.

MANFRED FÜLLSACK

NEIGHBORHOODS AND SOCIAL SECURITY

AN AGENT-BASED EXPERIMENT ON THE EMERGENCE OF COMMON GOODS

ABSTRACT

The paper investigates the emergence and probability of cooperation in repeated common-good-games in different network topologies. A multi-agent model is presented with agents interpreted as regularly and precariously employed workers which contribute with different probabilities to the common good of an income maintenance system. Agents memorize success and failure of investments and change probabilities in regard to their experiences. Differently structured neighborhoods, analytically seized in form of network topologies, support and constrain probability dynamics. Results indicate that the conditions supporting the emergence of cooperation might be sub-optimal for the maintenance of cooperation and vice versa.

INTRODUCTION

Income maintenance systems as deployed in European welfare states can be distinguished with respect to the significance they ascribe to non-contributory benefits in the case of unemployment.[1] Whereas in countries like France or Germany premium-based insurance systems prevail over non-contributory benefits, Scandinavian countries rather tend towards tax-based redistribution systems with less emphasis on insurance. Arguments for the one or the other system are manifold and discussions fill libraries. In times of dissolving standard employment relations and growing uncertainty in regard to labor curricula and careers, non-contributory benefits seem to gain reason. With rising numbers of unemployed school leavers or single parents falling out of the reach of insurances, redistribution systems that reallocate means not just *individually* (that is,

---

1 Richard M. Titmuss, *The Relationship between Social Security Programmes and Social Service Benefits: An Overview, Commitment to Welfare* (London: Allen and Unwin 1968). Robert Walker, *Social Security and Welfare: Concepts and Comparisons* (Buckingham: Open University Press, 2004). Jane Millar, ed., *Understanding Social Security: Issues for Policy and Practice.* 2nd edition (Bristol: Policy Press, 2009).

from a period in one's private life with employment to another period without employment), but *socially* (that is, from society members with employment to members without), appear to guarantee more security in terms of income maintenance.

Socially redistributing welfare systems, however, are subject to a typical problem which is characteristic for common goods.[2] By guaranteeing income irrespective of employment, such systems are prone to *free riding*. At least some beneficiaries might not care for employment if they can make a living on welfare.[3] The problem here is not so much the lone free rider on his own. Game-theoretic experiments indicate that the problem rather increases with the ones that might follow.[4] Free-riding can be contagious. The free riding of some diminishes the contribution-probability of others and so induces a downward spiral which eventually destroys the common good. In other words, the possibility to free-ride exposes socially redistributing welfare systems to a counter-current which impairs their attractiveness.

However, these dynamics might unfold in the other direction as well. The decrease of contribution-probabilities might be reverted by unprofitable experiences with overall non-cooperation.[5] If nobody contributes, the common good of a redistributing income maintenance system cannot be maintained. Nobody gains. But in consequence, exactly this negative experience might reinvigorate cooperation.

Analytically, two causes for this reinvigoration can be distinguished. As with the decline of contribution-probabilities, its increase as well seems to be supported by, on the one hand, individual strategic adaptation in regard to the payoff of a decision (i.e. learning), and, on the other hand, by the influence of others (i.e. other-regarding preference).

As experiments with networked common-good games[6] indicate, the influence of others is essential for the dynamics of contribution-probabilities in social dilemmas.

---

2 Mancur Jr. Olson, *The Logic of Collective Action: Public Goods and the Theory of Goods* (Cambridge Mass.: Harvard University Press, 1965).

3 Robert Moffitt, "Incentive Effects of the U.S. Welfare System: A Eeview," *Journal of Economic Literature* 30 (1, 1992), p. 1–61. Walker, *Social Security*.

4 Ernst Fehr and Simon Gächter, "Cooperation and Punishment in Public Goods Experiments," *American Economic Review* 90 (2000), p. 980. Ernst Fehr and Simon Gächter, "Altruistic Punishment in Humans," *Nature* 415 (2002), p. 137–140.

5 Natalie S. Glance and Bernado A. Huberman, "The Outbreak of Cooperation," *Journal of Mathematical Sociology* 17, 4 (1993), p. 281–302. Natalie S. Glance and Bernado A. Huberman, "The Dynamics of Social Dilemmas," *Scientific American* 3 (1994), p. 76–81.

6 Manfred Füllsack, "Evolving Networks of Cooperation. Experiments with Repeated Public Good Games and Evolutionary Computation," *Networking Networks. Origins, Applications, Experiments*, ed. Man-

The ways these influences spread through neighborhoods can be crucial for the fate of the common good. In order to gain insight into these paths of dissemination, I deployed an agent-based model (ABM) that positions agents in structured neighborhoods represented by different network topologies and confronts them with each other in repeated common-good games. Results give reason to assume a fundamental difference between the organizational structure that optimally supports the increase of cooperation-probability and the kind of structure that optimally fosters maintenance of cooperation once it has been achieved. In the following I will (1) shortly describe the social dilemma and the general dynamics of cooperation probability in common-good games. I will then (2) introduce the mentioned agent-based model in its basic, that is, not-networked form and present some results. I will briefly discuss some network aspects (3), deploy the model in various network topologies (4+5) and discuss respective results (6).

1. THE REAL TRAGEDY OF THE COMMONS

Free riding is a well known problem in so called social dilemmas,[7] that is, in situations in which individual and social benefits differ and individuals cannot easily be precluded from misusing or exploiting social achievements.[8] An illustrative theoretical example for this dilemma provides the so called Common Good-game in which non-contribution is individually profitable as long as it remains a minority behavior.[9] If the game is repeated, however, chances are high that the profits of the minority induces the majority to follow and in doing so to destroy its effects. In the end, non-contribution (*defection*) does not

---

fred Füllsack (Vienna: Turia & Kant, 2013).

7 Brian Barry and Russell Hardin, eds., *Rational Man and Irrational Society?* (Beverly Hills, Calif.: Sage, 1982).

8 These achievements are said to have the feature of *non-excludability from consumption*, meaning that users cannot, with justifiable costs, be kept from consuming. Doing so entails a control system which tends to get complex in the course of controlling. Compare Ross Ashby, *Introduction to Cybernetics* (London: Chapman and Hall, 1956).

9 Consider the following situation: three people get 20 Euros each with the invitation to invest all or a part or nothing of it into a common pool. Whatever sum will be in the pool after investment will be doubled and distributed equally among the players. If all players for example invest 20 Euros, each of them gets a final pay-off of 40 Euros. However, if one player invests, say, just 5 Euros, while the others still invest 20, the outcome is different. The pool will contain 45 Euros. Doubled this will be 90 which provides each of the players with 30 Euros. For the two 20-Euro-investors this is their final pay-off. For the 5-Euro-investor however, the 30 Euros add up with the 15 he didn't invest to 45, which is definitely higher than the 40 he might have received when investing all of the money.

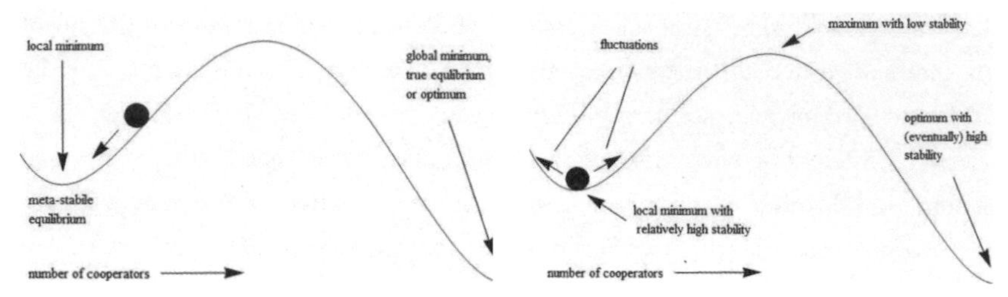

Fig. 1: Dynamics of the repeated common-good-game, pictured as a ball on a surface with two equilibrium solutions (*valleys*) separated by a mountain of unstable investment constellations. Since the solution to the social dilemma at the local equilibrium is suboptimal, there might be fluctuations in the position of the ball (right image) which eventually can become large enough to push the ball over the maximum on to the global equilibrium marking the optimal solution.

profit anybody, but behavior stays firmly put at the sub-optimal Nash-equilibrium. Nobody is willing to contribute because nobody else is willing either. The real "Tragedy of the Commons"[10] thus, is its self-accentuating dynamic.

The ABM-model described in Section 2 faces this problem in the way Natalie Glance and Bernado Huberman[11] suggested. According to them, the dynamics of a social dilemma when repeated can be pictured as the behavior of a ball on a rugged surface with valleys and peaks. Initially, this ball is placed on a random point that corresponds to a tentative first round of the common-good-game with most players investing but some deciding to free ride. When repeating the game, due to its contagiousness the probability of free-riding increases. Players copy each other's behavior and investment melts down to zero. The ball on the surface follows gravity and rolls to a valley, a local minimum that marks the Nash-equilibrium of overall zero-investment. Nobody gains from investing as long as nobody else invests. The ball stays put in the valley (Fig 1.).

This equilibrium, however, is suboptimal and everybody knows it. If the majority would invest they could profit more from the game. So obviously, there is a second valley on the surface which marks a more beneficial outcome with everybody profiting from the common good. Unfortunately, between these two valleys there is a high mountain of pretty unstable constellations consisting of different investments which all seem to push the dynamics towards the suboptimal equilibrium.

At this equilibrium, however, nobody profits. Therefore, it seems likely that with time all players will become aware of its sub-optimality. Unsatisfied with what they

---

10 Garrett Hardin, "The Tragedy of the Commons," *Science* 162 (1968), p. 1243–1248.

11 Glance, "Outbreak".

have, they might start tentative search actions. They might deviate temporarily from zero-investment, initially of course falling back to it every time they are ratted again. But if attempts are numerous and driven by unprofitable experiences with zero-investments, the probability to overcome the mountain behind which the optimal solution is hidden might rise.

2. THE MULTI-AGENT MODEL

In order to explore this possibility in more detail, I exposed a population of computer-generated software-agents to different experiences with income-certainty and varied their capacities to memorize these experiences.

In this model[12] a population of *p* agents is divided into two groups, *a* and *b*, with different certainties (i.e. probabilities) of receiving an income *i* from employment. Group *a* is regarded as regularly employed with high income-certainty c_a. Group *b* is seen as precariously employed with low income-certainty c_b. Agents receive incomes proportional to their income-certainty, with certainty being compared to a random real number between 0 and 1 and incomes paid whenever this number is smaller than certainty.

Agents face a level *n*, interpreted as minimum subsistence level, above which incomes accumulate in the form of wealth *w*.[13] If *w* exceeds *n*, agents can contribute to a social security fund *s* from which means *d* (for *dividend*) are distributed to the ones not getting any income in the current round (p_{ni} for *agent with no income*), with $d = s / p_{ni}$. Contributions are seen as investments *inv* into the common good of socially financed welfare *s*.

Precariously employed agents start out with an assumed *a priori* experience with being dependent on welfare, and therefore are more likely to contribute to it than regular employees. However, contribution for them is possible only if their wealth *w* exceeds the subsistence level *n*. In both groups, probability and amount of contributions depend on an initially ascribed cooperation probability (*coop*) which is low for regular (group *a*) and high for precarious employees (group *b*). To keep things simple, cooperation probabilities are taken as the complement of income-certainty *c*, that is: $coop = 1 - c$. (An agent contributes whenever a randomly drawn real number between 0 and 1 is lower

—

12 Generated with Netlogo: Uri Wilensky, *NetLogo*. http://ccl.northwestern.edu/netlogo/. (Center for Connected Learning and Computer-Based Modeling: Northwestern University, Evanston, IL. 1999).
13 Wealth can become negative. This influences the possibility of cooperation, but not the existence of agents.

than *coop*). The amount of investment then is determined as $inv = coop * (i - n)$. In other words, agents are not obliged to pay any predetermined amount of contribution as in institutionalized social security systems. Welfare here is not a step-level common good.[14]

Whenever agents do not receive income in a round of the game – be it due to the low income-certainty of being precariously employed or due to the relatively small residual uncertainty that regular employees face as well – they memorize this in the form of a binary input *e* (*experience*) into an income-memory – technically a list of length *ml*, with $e = 1$ for *no income* and $e = 0$ otherwise. Whenever they then get a dividend higher than zero from welfare, they memorize this as well in the form of a binary input into another list of length *ml*, regarded as welfare-memory (1 for *positive dividend*, 0 otherwise).

These memorized experiences then determine their willingness to contribute and the level of investments in welfare. That is, negative income experiences and positive welfare experiences raise the cooperation probability *coop* and thus the amount of investment. *coop* grows exponentially with a squared tenth of the sum of the respective experiences:

$$\text{coop}_{t+1} = \text{coop}_t + (0.1 * \textstyle\sum_0^{ml} e)^2.$$

Additionally however, cooperation probability is influenced by the number of contributors and non-contributors in the population. Each non-contributor ($inv = 0$) in the population diminishes cooperation probability by a per mill and each contributor ($inv >= 0$) increases it by a per mill.

When running the model with $p = 50$, $a = 45$ (90% of p, hence $b = 5$), $c_a = 0.9$, $c_b = 0.1$, $i = 10$, $n = 5$, the dynamics of cooperation-probabilities, that is the development of the agent's *willingness* to contribute to welfare (and, due to its relatedness, the amount of their contribution) strongly depend on memory-size *ml* (which is equal for both lists). With $ml < 6$, the mean cooperation-probability of agents rapidly falls to below 5% and stays put at this level (left plot in Fig. 2). The resulting security fund *s* does not get big enough to guarantee incomes above subsistence level *n* to all agents. About 20% of the population remain with an income $< n$. With a memory-length $6 <= ml < 9$, cooperation-probabilities at first also fall back to about 2% but then rise up to varying levels $< 30\%$ and fluctuate around these levels. Still not all agents can be endowed with

---

14 However, respective experiments with a fixed level of contribution did not result in significant differences.

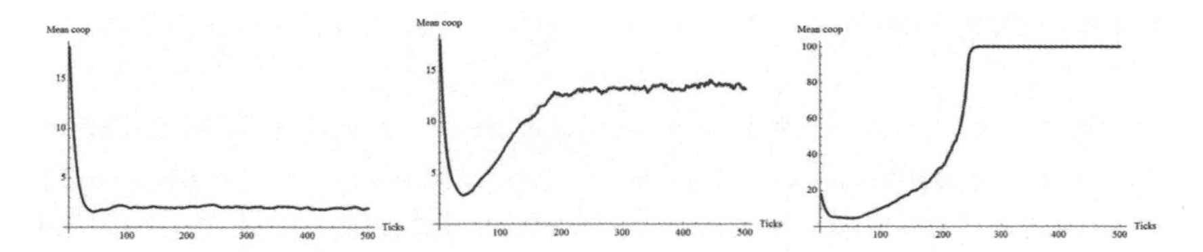

Fig. 2: Development of mean cooperation-probability in regard to memory capacities, with *ml* = 4, *ml* = 7 and *ml* = 10, from left to right.

incomes >= *n*. Eventually, starting from *ml* = 9, cooperation-probability – again after a rather sharp initial slump – rises slowly but steadily to about 40% and then due to feedback starts to accelerate upwards to 100% (right plot in Fig. 2). In this case, all agents are unfailingly endowed with means > *n*.

In this latter case, the dynamics of cooperation-probability follow a *Rapid Phase Transition*,[15] indicating that dynamics are no longer driven by *individual* experiences (and thus are not dependent on memory-length), but by *social influence*. The number of cooperators in the population enforces itself by copying behavior. The population exhibits aspects of herding.[16] The rise of cooperation-probability accelerates and cooperation *locks in*.[17]

The effect of this lock-in then shows in the fact that the stable contribution behavior of agents, once established, does not vanish if memory-length is reduced while running the model. Even without any memory at all, that is, with *ml* = 0, cooperation remains firmly at 100% if it once has been established. The mutual influence of contributors suffices to stabilize this kind of welfare system.[18]

—

15 Stuart Kauffman, *At Home in the Universe: The Search for Laws of Self-Organization and Complexity* (Oxford/USA: Oxford University Press, 1996). György Szabo and Christof Hauert, "Phase Transitions and Volunteering in Spatial Public Goods Games," *Physical Review Letters* 89, 118101 (2002).
16 William D. Hamilton, "Geometry for the Selfish Herd," *Journal of Theoretical Biology* 31 (1971), p. 295–311. Robert Keith Sawyer, *Social Emergence: Societies As Complex Systems* (Cambridge, Mass.: Cambridge University Press, 2005).
17 W. Brian Arthur, "Competing Technologies, Increasing Returns, and Lock-In by Historical Events," *Economic Journal* 99 (1989): p 116–131.
18 For similar stabilization effects see: Timur Kuran, *Private Truths, Public Lies. The Social Consequences of Preference Falsification.* (Cambridge, Mass.: Harvard University Press, 1995). Damon Centola, Robb Willer, and Michael W. Macy, "The Emperor's Dilemma. A Computational Model of Self-Enforcing Norms," *American Journal of Sociology* 110 (4, 2005), p. 1009–1040.

3. NEIGHBORHOODS

The behavior of the system changes when agents are arranged in networks and influenced not by the overall amount of free-riders or contributors in the population, but by the respective number among link-neighbors (that is, among agents to which they are connected with a link, cf. Fig. 3).

Basically, the scenario described in section 1 can be interpreted as a neighborhood with all neighbors being equally distanced to each other. Technically, this conforms to a complete network with equal distribution of links among agents. This distribution, however, is highly unlikely in real social conditions. Due to capacity limitations in maintaining social contacts,[19] social agents are more likely to organize in structured neighborhoods, for example in so-called Small World Networks[20] which consist of several rather densely connected cliques or local neighborhoods and some rather rare long-range connections between them. Technically, respective networks are specified as having a high Clustering Coefficient (*cc*) and a rather low Average Path Length (*apl*).[21] They are known for their efficiency in many fields of networked instances, such as synchronizing coupled phase oscillators, disease spreading, information diffusion etc.[22]

Small World Networks can exhibit *scale-freeness* in various amounts, meaning that their degree-distribution approaches a so-called *power law* distribution with a small number of nodes possessing a large number of links and a large number of nodes having a smallnumber of links. Mathematically, an ideal *power law* distribution can be obtained with the simple function $y = x^{-1}$. More generally, scale-freeness can be approached by

---

19 Robin I.M. Dunbar, "Neocortex Size as a Constraint on Group Size in Primates," *Journal of Human Evolution* 22 (1992), p. 469–493.

20 Stanley Milgram, "The Small World Problem," *Psychology Today* 1 (1961), p. 60–67. Duncan Watts, *Small Worlds. The Dynamics of Networks Between Order and Randomness*, Princeton Studies in Complexity (Princeton: Princeton University Press, 2004); Peter Csermely, *Weak Links. The Universal Key to the Stability of Networks and Complex Systems* (Berlin: Springer, 2009).

21 The Clustering Coefficient (*cc*) indicates the average of the local connectedness of the nodes, that is, the average relation of actual to possible connections of each agent, or in other words, their cliquishness. And the Average Path Length (*apl*) indicates the average number of steps necessary to reach one agent from any other agent in the network. As Duncan Watts and Steven Strogatz (1998) have shown, *cc* and *apl* do not develop conjointly when the probability of what they called "rewiring" is changed. *Apl* might decline faster then *cc* when the rewiring probability p rises, implying a p in the interval {0, 1} with maximal difference of *apl* and *cc*. Duncan J. Watts and Steven H. Strogatz, "Collective Dynamics of 'Small-World' Networks," *Nature* 393 (1998), p. 440–442.

22 Steven H. Strogatz, *Sync. How Order emerges from Chaos in the Universe, Nature and Daily Life* (New York: Hyperion, 2003).

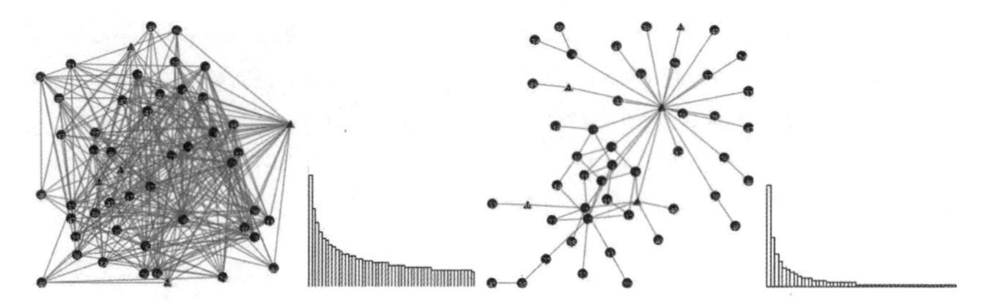

Fig. 3: Agent networks with corresponding link distribution plots, left with $\alpha = 0.5$ (*apl* = 1.82 and *cc* = 0.314), right with $\alpha = 1$ (*apl* = 3.003 and *cc* = 0.057). *Regularly employed* agents are shown as black circles, *precariously employed* as triangles (positions here are random).

arranging the links of the network according to the factor $p_i^{-\alpha}$, with p_i indicating the i-th member of the population and α being a real number, here in the interval {0, 1}.[23] In such an arrangement, the i-th agent should have[24] $i^{-\alpha}$ links. With this technique, an α-level of 0 specifies a completely connected network corresponding to the population structure in scenario 1, that is, to a society in which each agent is connected to each other and the amount of free-riders or contributors among link-neighbors therefore equals the overall amount in the population. Networks with α-levels > 0 exhibit scale-free degree distributions. An α-level of 1 for instance, specifies a network in which agents are connected to a share of other agents which corresponds to the reciprocal of their position in an agent's list. On average therefore, the i-th agent is connected to $i^{-1}*(p-1)$ other agents. The distribution of relations among neighbors is skewed. The following figure shows two agent-networks with different α-levels that were generated with this technique.

At times, networks in this context are interpreted in terms of political structures, with α close to 0 resembling democratic (or heterarchic) structures which are flexible, adaptive to change and good in coping with stress, but face high transaction costs and

---

23 Networks with $\alpha > 1$ tend to disperse, that is, to consist of unconnected parts. Therefore they are not considered in this paper.

24 When wiring the agents not randomly, but one by one with respect to $i^{-\alpha}$, agents who have been linked with the first agents might have already too many links at the time their turn comes up to be wired. The algorithm deployed here for this kind of wiring therefore wires all agents and subsequently checks whether their link-degree really corresponds to their list position. In this way, depending on α, the actual degree of p_i differs from $i^{-}\alpha$. In total, however, degree-distribution approaches a *Power law* distribution as Fig. 4 indicates.

clusterization, all due to a rich ecosystem of "weak links".[25] Networks with α around and greater than 1, on the other hand, stand out with a high percentage of "strong links" and resemble rather centralistic hierarchical structures which are efficient, stable and reactive in the case of emergencies, but tend to exclude valuable information and thus behave rather conservatively and unsusceptible to innovation.[26]

4. THE NETWORKED MODEL

Running the networked version of the model with parameters as described in section 1 shows interesting results. In general, the networked population needs significantly smaller memories to generate overall cooperation. The following table relates different α-level-networks to the minimum memory-length and the number of steps needed to reach 100% cooperation-probability. The necessary memory-length (column 2) strikingly decreases with the increase of α (column 1). But the number of steps needed to reach 100% *coop* (column 3) vigorously increases.

α	*ml*	*steps*	*apl*	*cc*	*links*
0.2	7	493	1.427	0.586	693
0.4	5	1013	1.696	0.363	387
0.6	4	1078	2.037	0.243	202
0.8	3	2010	2.348	0.398	112
1.0	2	6750	3.062	0.1	70

Table 1. Minimum memory-lengths (*ml*) at which 100% cooperation-probability can be reached. *Steps* indicate the mean number of iterations (rounded from 10 consecutive runs) needed to reach 100% mean *coop* (*apl*, *cc* and the number of *links* in the network (also as rounded means) are given for orientation).

The following plots relate mean cooperation-probabilities (*y*-axes), reached after 2500 steps, to increasing α-levels (*x*-axes) in agent populations with (from left to right) memory-lengths of $ml = 3$, $ml = 2$ and $ml = 1$.

Unfailingly, 100% cooperation-probability is reached with memory-lengths $>= 3$ and α-levels $>= 0.6$ (that is, when mean *coop* grows beyond 35%). However, the *coop*-level, at which an endowment of all unemployed with means that cover their needs

---

25 Eric L. Berlow, "Strong Effects of Weak Interactions in Ecological Communities," *Nature* 398 (1999), p. 330–334; Csermely, *Weak Links*, p. 3.
26 Csermely, *Weak Links*, p. 257f.

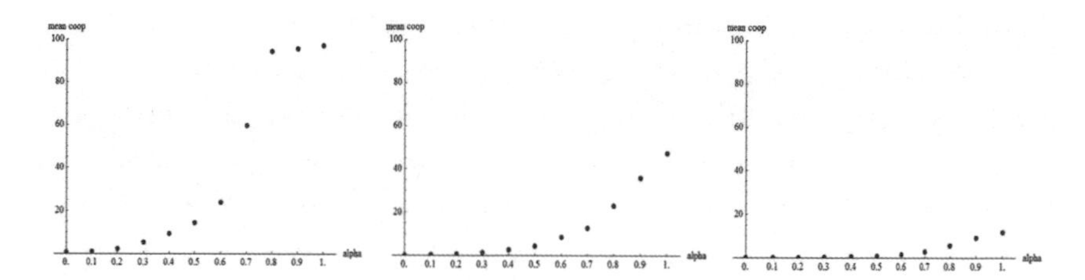

Fig. 4: Mean cooperation probabilities, reached after 2500 steps, in relation to α-levels, left with *ml* = 3, middle with *ml* = 2 and right with *ml* = 1.

seems to become feasible, is reached much earlier. A sort of *tipping point* demarcates the level of *coop* at which the dynamics of contribution-behavior remain beneath the level of guaranteeing subsistence to all unemployed from a level at which the probability keeps steadily rising, although with differing pace. With the parameters given above, this *tipping point* amounts to about 35% mean cooperation-probability. It is the point at which the social influence starts to outweigh the effect of the memorized experiences (or in other words, at which the lock-in takes effect). The following table correlates this point with α-levels, memory-lengths and (rounded) mean steps needed to reach it.

α	*"tipping point" (mean coop)*	*ml*	*steps*
0.2	34.16	7	250
0.4	32.78	5	303
0.6	36.78	4	331
0.8	35.03	3	431
1.0	32.43	2	842

Table 2. *Tipping points* (mean of 10 runs) from which cooperation probabilities keep rising to 100% in relation to memory-lengths and steps needed to reach them.

These results[27] suggest that particular neighborhoods are more supportive than others for the generation of cooperation in terms of willingness to contribute to a socially distributing income maintenance system. Rather centralistic and hierarchical structures, that is, such with α around 1, seem to be gainful when a welfare system is established in a society where most regular employees seldom face negative income

---

27 These results were reproduced with parameters $p = 50$, $a = 75, 80, 85, 90$ and 95% of p and b respectively, $c_a = 75, 80, 85, 90$ and 95, $c_b = 5, 10, 15, 20$ and 25, $i = 10$, $n = 5$. From $a < 75\%$ (with b, c_a and c_b respectively reduced) the model shows slightly different dynamics which will be discussed elsewhere.

experiences (that is, agents possess low *ml*). On the other hand, a rather de-centralised or heterarchically organized society (with low α) seems to foster the emergence of cooperation under conditions where experiences with dependence on welfare is rather common even among the regularly employed (that is, if agents have high *ml*). Expressed a bit more pointedly, infrequent majority dependence on welfare seems to need rather *autocratic* social structures in order to have society-wide income maintenance emerging. With more frequent overall dependence on welfare, rather *democratic* structures suffice.

5. THE STABILITY OF NEIGHBORHOOD COOPERATION

Admittedly so far these results are only moderately exiting. They correspond to what one intuitively might have expected. However, it should be mentioned of course, that speaking of *autocracy* in this context does not imply any centralistic plan or top-down instruction to cooperate. The emergence of cooperation proceeds bottom-up and self-organized, albeit with different paths to disseminate. However, the behavior of the model becomes more interesting if we ask for the stability of cooperation once the common good of society-wide income maintenance has been established (that is, once 100% cooperation-probability has been reached). Different neighborhoods foster stable cooperation to varying extents. And what is more, important topologies that prove efficient in generating cooperation seem to be less suitable for maintaining it.

In order to demonstrate this, the model was kept running once 100% *coop* was reached and memory-length was reduced to zero. As mentioned in Section 1, in this case cooperation depends solely on social interactions and no longer on experiences. In networks with $\alpha <= 0.5$, cooperation-probability remains unfailingly at 100%. Obviously, any perturbations by non-contributors can be successfully dispersed. Free-riders have influence, but their effect is bypassed via redundant connections. With $0.5 > \alpha < 0.8$, *mean coop* starts to diminish slightly when memories are switched off, but stabilizes at still high levels – about 95%. At $\alpha >= 0.8$ *mean coop* begins to cave in more dramatically. At these α-levels the stability of cooperation starts to depend very much on the position of non-contributors in the network. Since few or no connections to their effects exist at this point, non-contributors – be they coincidental free-riders or precariously employed who cannot afford cooperation – can have a quite important influence on the stability of cooperation. The following figure shows two $\alpha = 1$-networks in which precariously-employed agents are centered, meaning that they are positioned as highly-linked net-

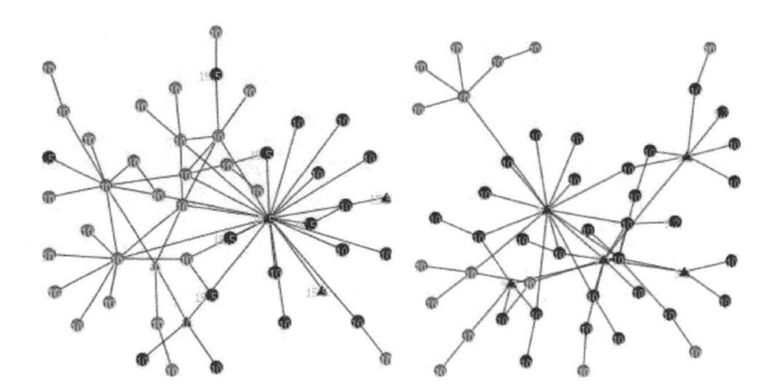

Fig. 5: Topologies and mean cooperation probabilities in networks with $\alpha = 1$, after reducing memory length to zero when reaching a *mean coop* of 100%. Left: one precariously employed (shown as red triangle) in centered position, Right: all (here 5) precariously employed in centered position. Contributing agents green, non-contributing red.

work hubs with high betweenness-centrality.[28] In the left network just one of them is centered, in the right all of them are. The plot beneath documents the corresponding decline of cooperation-probability.

6. DISCUSSION

As the right plot in Fig. 5 suggests, cooperation probability can fall back to below subsistence-guaranteeing levels in neighborhoods with most precariously employed agents being highly linked. Income maintenance and therewith social security can be forfeited again. The ball of the common good crests the mountain of unlikely investment constellations in the opposite direction and rolls back towards the suboptimal Nash equilibrium of no-one investing because no-one else is investing either.

However, this is true only for rather centralized hierarchical topologies with little redundancy in social interactions. In networks with low α-levels, that is with more possibilities to disperse social influence, cooperation remains widely stable. In such neighborhoods, however, cooperation does not seem to find optimal conditions to emerge in the first place. It looks as if the generation and the maintenance of cooperation blossom in different ecologies. If this result can be generalized – for example into the realms of

—

28 Betweenness-centrality indicates the number of shortest paths on which a node is positioned in respect to all other nodes.

organization theory – one might be inspired to speculate about ways to rearrange society (or the staff of a department, the employees of a firm) in respect to whether the emergence or rather the stability of cooperation is at stake. One might also muse about the question whether democracies in their organizational structure provide just the right *in-between* state to foster both dynamics. As Peter Csermely[29] for instance points out, modern democracies stand out with a set of stable hierarchical, centralized structures with rather strong links and little redundancy while at the same time exhibiting a wide variety of rather weak and permanently fluctuating links which sustain the necessary social influence for stabilizing common goods even without all individuals benefiting.

Another interesting aspect arises from the question whether the emergence of cooperation might influence society in a way that feeds back on its link-structure, that is, on its α-value. Could the successful emergence of cooperation change the conditions for its stability? The model, as introduced in this paper, is too simple to investigate this possibility. But in more realistic scenarios one might reason that the emergence of a society-wide income maintenance system could provide a kind of fundamental wellbeing that enables and fosters a wide range of new social and cultural activities and interactions – activities and interactions which would not be possible with a part of society permanently struggling on the edge of subsistence. In terms of network theory, these new interactions would be observed as links that increase the biodiversity of society and therewith enhance its cohesion. Technically, these additional links diminish the original α-level and thereby might increase the stability of cooperation. Holistically seen, cooperation could increase its own probability, driving social dynamics from rather centralistic hierarchical neighborhoods which foster emerging cooperation towards more de-central heterarchical neighborhoods which promote its stability.

—

29 Csermely, *Weak Links*.

SEBASTIAN GIESSMANN

TOWARDS A MEDIA HISTORY OF THE CREDIT CARD

ABSTRACT

The paper sketches out a praxeological history of the credit card, with an emphasis on its mediating qualities. Temporal borrowing via credit card is regarded as a highly dynamic neighborhood technology that produces emergent topologies through distributed mobile payments. The formative years of the American credit card (1950–1975) are analyzed along with approaches from ethnomethodology (Harold Garfinkel) and actor network theory (Michel Callon). Cooperative media practices are key to understanding the relation between a neighborhood technology and its social networks, including (1) Dining, traveling and charging, (2) Accounting for trust and credit, (3) Mass mailing and advertising new ways of payment, (4) Building "co-opetitive" platforms for networks and (5) the digital momentum of credit cards.

INTRODUCTION

It is a rather easy task to picture "giving credit" as an everyday neighborhood social practice. "Could I borrow something?" might be easier to respond to in a cooperative way than "Can I get a temporary loan?". Nonetheless, it is hard not to help someone who is living close to you. It is a bit harder to imagine that credit cards have been a neighborhood technology right from the beginning, though. But the small plastic machines have been and remain tied to local territories and sites of exchange, even in a highly globalized sphere of digital transactions. Credit cards are a highly mobile medium of cooperation, yet their enabling infrastructural architectures – be they organizational or technical – develop by user interaction, national markets and in correspondence to the legal system. Within the history of credit and debit cards, a rather remarkable transformation of proximity and distance took place in its mediated neighborhoods. This article attempts to trace them.[1]

1 I am highly indebted to Erhard Schüttpelz, Jens Schröter, and the Working Group "Media of Cooperation" at University of Siegen for the ongoing discussion.

My contribution therefore has a rather humble target – following the media practices that have made the American credit card an essential part of our digitized media culture. Rather than taking today's Visa, American Express, and Master Card networks for granted I am going to return to the initial stages that turned "paying with plastic"[2] into an everyday practice. The list of practices is by no means going to be complete, but it attempts to integrate what ethnomethodologists have understood as "reflexivity" of everyday actions. Rather than understanding acts of payment as a routine operation that is becoming a somewhat tacit action, ethnomethodology would insist that even the most common situations rely on a reflexivity of the actors, which is intelligible by their accounts of "what is going on while it is going on." Actions have to be made accountable through indexical expressions in every interactional setting, as Harold Garfinkel has shown in his seminal works.[3] One of Garfinkel's methodological interventions was his artful method of listing everyday practices. This should not be understood as a merely arbitrary textual device. Rather, it provides a good way of reorientating media history towards the interactional und situational production of all kinds of media. So out of a possibly endless Garfinkelian list of practices, the following are questioned regarding the "accountability" of credit cards: (1) Dining, traveling, and charging, (2) Accounting for trust and credit, (3) Mass mailing and advertising new ways of payment, (4) Building "co-opetitive" platforms for networks and (5) the digital momentum of credit cards.

Listing the interactional practices of credit card usage does not artificially separate their indivisible meshwork. Rather, credit cards "in action" create dynamic topologies that push the payment infrastructures with their necessarily bureaucratic regimes to their boundaries. They establish a "mediality of proximity,"[4] in which your account is following you to wherever a credit sign might appear. These practices evolved mostly between 1950 and 1975, although some of them like the "charge-a-plates" have a longer history, as they were already used by oil companies and department stores before World War II. On another note, this article is not going to dig deeper into the world history of credit, capitalism and its media (yet). Nonetheless, Fernand Braudel's works on the *longue durée* of economic practices as operational chains form an epistemological

—

2 David S. Evans and Richard Schmalensee, *Paying with Plastic. The Digital Revolution in Buying and Borrowing*. 2nd edition. (Cambridge, Mass.; London: MIT Press, 2005).

3 Harold Garfinkel, *Studies in Ethnomethodology* (Englewood Cliffs, NJ: Prentice Hall, 1967).

4 See Pablo Abend, Tobias Haupts and Claudia Müller, eds., *Medialität der Nähe. Situationen – Praktiken – Diskurse* (Bielefeld: transcript, 2012).

background here, as well as recent ethnological approaches like David Graeber's work on debt and value, and Michael Mann's analysis on the sources of social power.[5]

There is a remarkable amount of literature available on historical aspects of the credit card, but its status as a now seemingly older social and bureaucratic medium of accountability and its pioneering digital infrastructures certainly deserves more attention. So far, David L. Stearn's excellent study of VISA's computational system's building has been one of the few books to move into that direction.[6] The following four steps "in the footprints of practices" shall add further to this approach and address how these practices relate to payment and credit platforms.

As we shall see, those platforms have developed into a prime mover of dynamic social networks. They even create their own statistical and predictive tools of "social mathematics," most notably through the scoring of creditworthiness by relating geocode, social stratification data, and personal payment practices. Payment and credit platforms are media neighborhoods in themselves, yet they evolve out of mobile local and regional practices of payment, both on the interactional and organizational level. For the individual, credit cards were marketed as a portable object of trust, giving national and global access to a bank account and thereby mimicking the situation of owing money in a concrete social neighborhood. Yet they also established a new, paradoxical form of organizational proximity between competing banks. They had to work cooperatively together, shaping an interoperational extended neighborhood of debt and payment. Economics has found a neologism for this new mode of capitalism by calling it "co-opetition," thus describing a mix of *co*operation and com*petition*. I will come back to this notion later on.

1. DINING, TRAVELING, AND CHARGING

In every good American tale of entrepreneurship there is at least one hero, and an initial magical scene of invention has to be told. So it does not come as a surprise that the story of Diners Card as a New York neighborhood technology has been told as an anecdote

---

5 Fernand Braudel, *Afterthoughts on Material Life and Capitalism* (Baltimore; London: Johns Hopkins University Press, 1977); David Graeber, *Debt. The First 5,000 years* (New York: Melville House, 2011); Michael Mann, *The Sources of Social Power. Globalizations, 1945–2011*. Vol. 4. (Cambridge: Cambridge University Press, 2013).

6 David L. Stearns, *Electronic Value Exchange. Origins of the VISA Payment System*, History of Computing. (London: Springer, 2011).

ever since the 1950s. By now, we know that Frank MacNamara did not conceive of the idea while lacking the cash (or the wallet) to pay for his business dinner at Manhattan-based *Major's Cabin Grill* in 1949. He did not call his wife by telephone to bring in the money, either. The concept rather came up as an everyday idea that MacNamara wanted to put to a test, together with Ralph Schneider as his business partner. In the end, that lovely New York story has proven to be a quite successful PR device that Matty Simons, the first press agent of the firm, has popularized.[7] The idea itself was not particularly new, as multiple narrations of Diners Club's inception have emphasized. But it managed to combine elements that had not been unified before.[8]

Customer cards had been around before World War II, and big department stores, oil companies, hotel franchises and airlines used them to attract and bind customers: "Following the war, a number of large retailers joined together to form cooperative card operations similar to those begun in the 1930s. In 1948, a number of major New York department stores, including Bloomingdale's, Arnold Constable, Franklin Simon, Gimbel's, and Saks, began a charga-plate group; the standard charga-plates they mailed to their customers were usable at any of the cooperating stores."[9]

Interestingly, MacNamara's and Schneider's attempt to create a Diners Club did not come in the form of a metal "charge-a-plate" but as a small booklet with a business card style "card" on top that would inform about the places of acceptance.[10] The entrepreneurs also approached selected potential customers directly, mailing unsolicited cards to "several thousand prominent businessmen with a letter revealing its wonders."[11] MacNamara, Schneider, and Simmons managed to sign up 42,000 cardholders in the first year of operation. Their customers were all willing to pay $18 for the privilege of paying with their Diners Club "card". The company managed to establish what is now called a multisided platform market,[12] starting off with fourteen New York restaurants

—

7 Stearns, *Electronic Value Exchange*, p. 13. The Wikipedia entry on Diners Club repeats variants of the heroic story, but omits Stearns' clarification. Compare https://en.wikipedia.org/wiki/Diners—Club—International (last accessed September 18, 2014).

8 Evans and Schmalensee, *Paying with Plastic*, p. 149.

9 Lewis Mandell, *The Credit Card Industry. A History* (Boston: Twayne, 1990), p. 25. Evans and Schmalensee: *Paying with Plastic*, p. 53.

10 Stearns, *Electronic Value Exchange*, p. 13.

11 Matty Simmons, *The Credit Card Catastrophe. The 20th Century Phenomenon that Changed the World* (New York: Barricade, 1995), p. 27.

12 Evans and Schmalensee, *Paying with Plastic*, p. 3, 6–8, 150–52. Compare Jean-Charles Rochet and Jean Tirole, "Platform Competition in Two-sided Markets," *Journal of the European Economic Association* 1.1 (2003), p. 990–1029.

which had become 330 U.S. restaurants, hotels and nightclubs at Diners Club's first anniversary.[13] Besides the annual member fee, it generated its revenue from the 7% average fee that the owners of those restaurants, hotels and nightclubs had to pay as their part of the bill.

Using credit cards by establishing them as a Travel & Entertainment (T&E) novelty certainly was a success story of the 1950s. It quickly caught the attraction of other contenders. Mostly, they were not able to repeat what Diners Club had managed to do. There was a whole number of unsuccessful attempts to install charge and credit programs in banking throughout the 1950s and 60s, e.g. the Chase Manhattan Charge Plan (CMCP).[14] And we should symmetrically name some more credit card ventures, many forgotten, some merged with Diners Club, and a few others on their way to a further economic revolution: National Credit Card, Inc. (1951), First National Bank of San Jose (1953), Trip-Charge (1955), Universal Travelcard (American Hotel Association, 1956), Esquire (1957), Duncan Hines (1957), Gourmet Magazine (1950s), Bank of America (California, 1958), American Express (1958), and Carte Blanche (by Hilton, 1958).[15] Only Bank of America's own Bank Americard, American Express and – partly – Carte Blanche were going to surpass Diners Club in a second wave in the 1960s. A swift computerization of information processing was an important part of this development, as we shall see. And, until as late as the 1980s, a tremendous amount of financial loss and low profit margins, too.[16] Losses and low profits stood side by side with new and unexpected cooperative organizational developments in banking.

There has been quite a lot of speculation on the reasons why consumer credit proved to be such a specialty development of the United States, leading up to a veritable "Credit Card Nation" after a fragmented highly local, "wild and innovative" growth of credit card usage.[17] Lewis Mandell, the first eminent historian of the credit card, has made the case that after World War II government programs such as worker's compensation, unemployment and disability insurance paved the way for a rising consumerism – it

---

13 Evans and Schmalensee, *Paying with Plastic*, p. 54.

14 Mandell, *The Credit Card Industry*, p. 30.

15 Evans and Schmalensee, *Paying with Plastic*, p. 55–56.

16 Evans and Schmalensee, *Paying with Plastic*, p. 238.

17 Robert D. Manning, *Credit Card Nation. The Consequences of America's Addiction to Credit* (New York: Perseus, 2000); Lizabeth Cohen, *A Consumer's Republic. The Politics of Mass Consumption in Postwar America* (New York: Vintage, 2003), p. 123–124; Christine Zumello, "The 'Everything Card' and Consumer Credit in the United States in the 1960s," *Business History Review* 85 (2011), p. 551–575, p. 574.

was government spending that set the mood for credit.[18] This Keynesian background, along with the rather stable Bretton Woods financial system, led to a hitherto unknown spread of consumerism in the Western welfare states. The US first and foremost benefited from its wartime economy and innovation in these "thirty glorious years" of capitalism, lasting from 1945 to 1975.[19] The American middle class turned to credit once the wartime savings were spent. Consumers asked for more credit – and they got it by a flourishing financial and real estate industry that made credit purchasing *the* currency of the so-called "Consumers' Republic."[20]

Digging deeper into the history of mentalities and economic exchange usually brings up both rural and urban neighborhood practices of installment and revolving credit that date back at least to the year 1800.[21] As David Stearns has convincingly argued, there is also both a political and infrastructural momentum in handling the growing usage and clearance of checks that parallels this development. Handling und regulating this constant flow of paperwork even played a role in the establishment of the US Federal Reserve System in 1913.[22] The American geographies of exchange with their long distances called for neighborhood technologies of mediated trust, especially once the mobility of a whole society grows like it did in the 19th and 20th century.[23] Suburban shopping centers and credit cards form two sides of a new coin.[24]

How can we step along those lengthening chains of mediations that create the dynamic and fluctuating topologies of social interaction, "interessement" and "enrolment" of the actors[25] and payment practices? And what makes the prestigious Diners Club or American Express card such an effective and attractive medium, one that is

---

18 Mandell, *The Credit Card Industry*, p. 22. The enormous long-term influence of government spending for a multitude of privately owned innovations in America has been shown repeatedly. Compare Kenneth Flamm, *Creating the Computer. Government, Industry, and High Technology* (Washington, D.C.: The Brookings Institution, 1988); Committee on Innovations in Computing et al., eds.: *Funding a Revolution. Government Support for Computing Research* (Washington: National Academy Press, 1999); Mann, *The Sources of Social Power. Globalizations*, 1945–2011, chap. 2–5.

19 Mann, *The Sources of Social Power. Globalizations*, 1945–2011, chap. 3, p. 46.

20 Cohen, *A Consumer's Republic*, p. 147.

21 Mandell, *The Credit Card Industry*, p. 14–17.

22 Stearns, *Electronic Value Exchange*, p. 2–6.

23 I do not agree with Mandell's argument that this explains the European reluctance toward credit cards, though. Compare Mandell, *The Credit Card Industry*, p. 16, 154.

24 Cohen, *A Consumer's Republic*, p. 282–283.

25 Michel Callon, "Some Elements of a Sociology of Translation: Domestication of the Scallops and the Fishermen of St Brieuc Bay," *Power, Action and Belief. A New Sociology of Knowledge?* Ed. John Law. (London; New York: Routledge, 1986), p. 196–233.

mobile, bureaucratic, and cooperative at once? Along with Harold Garfinkel and Michel Callon, some preliminary answers might be given.

2. ACCOUNTING FOR TRUST AND CREDIT

The production of "accountability" is a multifold process, or as Garfinkel would put it himself, a contingent "ongoing accomplishment" in common situations.[26] Making things accountable includes "the activities whereby members produce and manage settings of organized everyday affairs [which] are identical with members' procedures for making those settings 'account-able'."[27] This includes a reflexivity of intertwined actions that rely on a high degree of social indexicality. Everyday reflexivity is based on explicit gestures and speech acts of showing and "telling the others," and tacit inter-corporeal ways of bodily interrelation.[28] Garfinkel uses also his artful way of listing practices to explain the modalities of "accountability": "In exactly the ways in which a setting is organized, it consists of methods whereby its members are provided with accounts of the setting as countable, storyable, proverbial, comparable, picturable, representable – i.e., accountable events."[29] What seems like a sociological pun does already include the formalities of institutional settings[30] and machine automatizations, e.g. of bank accounting. All bureaucratic media, especially files and forms in the case of money and credit, exist to produce a rather rigid infrastructural form of "accountability." By setting the standards for what is detectable, countable, comparable, and representable they mediate social relations as well as producing them. Making oneself "accountable" in such mixed realities combines the "rigid" bureaucratic media and the performativity of interaction between people, objects, signs, and institutions – including all possible conflicts of "norms, tasks and troubles."[31]

—

26 Garfinkel, *Studies in Ethnomethodology*, p. 33.

27 Garfinkel, *Studies in Ethnomethodology*, p. 1.

28 See Christian Meyer, Jürgen Streeck and Jordan Scott: "Intercorporeality: Beyond the Body. An Introduction," *Intercorporeality: Beyond the Body*. Ed. Christian Meyer, Jürgen Streeck and Jordan Scott (Oxford; New York: Oxford University Press, 2015) (forthcoming).

29 Garfinkel, *Studies in Ethnomethodology*, p. 34.

30 Compare Harold Garfinkel, "Good organizational reasons for 'bad' clinical records," *Studies in Ethnomethodology* (Englewood Cliffs, NJ: Prentice Hall, 1967), p. 186–207.

31 Garfinkel, *Studies in Ethnomethodology*, p. 33.

Interestingly, the credit card mobilizes the usual bureaucratic and economic procedures to an unknown extent. A highly localized American bank account now extends way beyond the rural and urban neighborhood, because it can be used at every place that accepts the new way of payment. This makes the credit card a cooperative social medium that is changing the geographies of shopping. It creates what ethnomethodologists call a "technology of accountability"[32] which is producing new portable neighborhoods that are able to lengthen chains of operations substantially. This spatial dynamic might not be new to capitalist modes of exchange, as one can learn from the seminal works of Fernand Braudel who has shown that the evolution of merchant capitalism in the Mediterranean and Europe has relied on long distance systems of trust.[33] In fact, paper-based transactions extended the geography of mercantile circulation and action at a distance, at least from the Middle Ages onward.[34] Just remember Marshall McLuhan's dictum on "Money: the poor man's credit card" in *Understanding Media:*

> Shortly before the event of paper money, the greatly increased volume of information movement in European newsletters and newspapers created the image and concept of National Credit. Such a corporate image of credit depended, then as now, on the fast and comprehensive information movement that we have taken for granted for two centuries and more. At that stage of the emergence of public credit, money assumed the further role of translating, not just local, but national stores of work from one culture to another.[35]

On a smaller level, this translation is getting repeated with the interconnection of personal bank accounts, mass consumer credit, networked systems of trust, and digital computing that is novel to American postwar culture. Initial success like the one Diners Club enjoyed early – even on an international scale – created massive back office problems.[36]

---

32 See also Lucy Suchman, "Technologies of Accountability. Of Lizards and Aeroplanes," *Technology in Working Order. Studies of Work, Interaction, and Technology*. Ed. Graham Button (London; New York: Routledge, 1993), p. 113–126.

33 Braudel: *Afterthoughts on Material Life and Capitalism.*

34 I am not able to elaborate on the point of form-constant paper technologies and circulating "immutable mobiles" extensively here. Please refer to Erhard Schüttpelz, "Die medientechnische Überlegenheit des Westens. Zur Geschichte und Geographie der immutable mobiles Bruno Latours," *Mediengeographie.* Ed. Jörg Döring and Tristan Thielmann (Bielefeld: transcript, 2009), p. 67–110.

35 Marshall McLuhan, *Understanding Media. The Extensions of Man. Critical Edition.* Ed. Terrence W. Gordon. 2nd ed. (Berkeley: Gingko Press, 2011), p. 192.

36 This is a reoccuring phenomenon in the history of cooperative media, as JoAnne Yates and Kjeld Schmidt have shown. Compare JoAnne Yates: "Evolving Information Use in Firms, 1850–1920. Ideology and

By now, it does not come as a surprise that American Express gained a significant advantage when it started using computers for its billing und bookkeeping in the early 1960s and surpassed Diners Club in 1970.[37] By then, it was a step that few banks and other businesses were taking in such a consequent manner, although many of them tried to do so. So while the impact of IT on credit card banking cannot be underestimated, it should also not be exaggerated – the social dynamics of the credit business are mostly external to computerization, yet they were pushing the technological development of cooperative digital computing resources. As historian of technology Thomas Haigh has shown, there was a wide gap between the American computing hypes and the actual adoption of new digital accounting practices.[38] This has to be kept in mind, because the laboratory situation of digital networking or "netting," as J.C.R. Licklider called it,[39] has brought up so many different localized approaches in the US between 1950 and 1975, that the successful endeavors enjoy a prestigious historical status. Just consider SAGE for the military purpose of missile defense, American Airlines' and IBM's SABRE for airline ticketing, or the now seemingly magic ARPANET for academic resource sharing.[40] Out of the early special purpose digital networks, the economic usages likely had the biggest impact for everyday use before the World Wide Web: mass consumer credit seemingly pushed forward the need for digital accounting and trusted transactions. But was this due to mere necessity in data processing, or is there a more specific relation between the history of credit, neighborhood "technologies of accountability," and the theory of (digitally) networked social media and their platforms? I want to come back to this question in the end.

—

Information Techniques and Technologies," *Information Acumen. The Understanding and Use of Knowledge in Modern Business.* Ed. Lisa Bud-Frierman (London; New York: Routledge, 1994); Kjeld Schmidt: *Cooperative Work and Coordinative Practices. Contributions to the Conceptual Foundations of Computer-Supported Cooperative Work (CSCW)* (London: Springer, 2011), chap. 11.

37 Stearns, *Electronic Value Exchange*, p. 17.

38 Thomas Haigh, "Technology, Information and Power. Managerial Technicians in Corporate America, 1917–2000" (Ph.D. diss. University of Pennsylvania, 2003), section III. "From Data to Information 1959–1975". "The question is not why a few companies successfully spent a fortune to push the state of the art. It is why so many more did not.," ibid., p. 323–324.

39 Joseph Carl Robnett Licklider Papers, MC 499.3, Correspondence 1958–1969. Massachusetts Institute of Technology. Institute Archives and Special Collections, Cambridge, Mass. Typoscripts, Memorandum to A. H. Eschenfelder (IBM), May 3 (*What IBM Should Do in the Field of Computer Networks*, p. 1) and May 10 1967 (*Burgeoning of Activity in the Field of Computer Networks*, p. 1).

40 See also Charles P. Bourne and Trudi Bellardo Hahn, eds., *A History of Online Information Services, 1963–1976* (Cambridge, Mass.; London: MIT Press, 2003).

3. MASS MAILING AND ADVERTISING NEW WAYS OF PAYMENT

In the mid-1960s, there is a peculiar change to be noticed concerning the social realities of credit card usage. Diners Club, American Express and a lot of other "go-it-alone" firms supplied a service that appealed to the white traveling salesman and the upper class, with accounts that primarily referred to this person alone. Throughout the 1960s and 1970s this social background was being transformed, making credit cards a payment technology that could pretty much be used by the whole middle class. Gender gaps were still highly present, most visibly in advertising,[41] yet credit by card soon extended its reach into whole families and company usage of corporate accounts. Actual discrimination of women persisted on a wide scale, though. Mortgage lenders and credit cards vendors like Carte Blanche and American Express required men to be the legal holder of credit accounts. But gender roles were also changing quite drastically in a short time: the overall "gendering of the 'consumer' shifted from women to couples, and at times to men alone."[42] Changes in the social stratification of credit did not come out of the blue, but responded to the dynamic unsolicited mass mailings of new cards that were addressed to pretty much everyone, including children and dogs.[43] This wild and highly dynamic phase quickly brought up issues of legal regulation, and it created a significant demand not just for market research but also for empirical studies of credit card use.

Within this context some details of marketing credit cards to the masses were analyzed swiftly by social research. The young Lewis Mandell prepared an extensive field survey for University of Michigan's Institute for Social Research in Ann Arbor, interviewing 3,880 "heads of households" in two waves in January, February, April and May of 1970.[44] Its carefully prepared forms and computed statistics already show some of the social biases and a stratification that was influential both for marketing and usage. Mandell is pretty straightforward in his brief narration:

> "Perhaps the most important explanation of why credit cards are found more frequently among higher income families is that credit cards have been marketed to these people from the beginning. [...] In fact, a great number of current card holders first became acquainted with credit cards as the result of unsolicited mailings of cards in the 1960's.

---

41 Zumello, "The 'Everything Card' and Consumer Credit in the United States in the 1960s".

42 Cohen, *A Consumer's Republic*, p. 147.

43 Mandell, *The Credit Card Industry*, p. 57.

44 Lewis Mandell: *Credit Card Use in the United States* (Ann Arbor: University of Michigan Press, 1972), p. 5.

> Recipients of bank cards, for example, were chosen from bank listings of persons with above minimum checking or saving account balances over time or persons who were reliable borrowers. The initial mass mailings were deemed necessary in order to get the operation started. A certain number of cards had to be put in circulation at once since the companies had high start-up costs and could not afford to wait the years it would take to build a demand for the cards."[45]

Within the credit card rush critics noted that only a fifth of the banks did actually really check the economic background of the addressees.[46] Hence, the anecdotes on dogs and children do not seem as unlikely as they sound, and the lax and overhasty mailing campaigns have been confirmed by oral history interviews,[47] and they were also part of contemporary cartoons.[48]

These actions were meant to make the case for signing up more merchants, so that both sides could be enrolled in the two-sided platform market. This marketing device can be traced back right to the beginning of what would later become Bank Americard (which subsequently turned into the Visa association in 1976). Bank of America ran its first market test in Fresno, California, based on the mass mailing technique in fall 1958.[49] Side effects of this strategy – namely, credit card theft, fraud and significant losses – had been known to the card business even before the Second World War. Nonetheless, the 1960s rise of the big co-opetitive networks of Bank Americard (the future Visa) and Master Charge (the future MasterCard) involved a race to the bottom. Who could get out more cards to more people in a shorter amount of time? "Network accidents"[50] like fraud were taken for granted by the banking industry, which multiplied its marketing efforts. Speeded-up mailings were one massive effort to differentiate the nearly identical services in payment and credit. On the other hand, branding and advertising attempted

---

45 Mandell, *Credit Card Use in the United States*, p. 12–13. See also Evans and Schmalensee, *Paying with Plastic*, p. 72–73.

46 Paul O'Neil, "A Little Gift from Your Friendly Banker. The mails bring credit cards to millions, opening new vistas from crime, chaos and comedy," LIFE 68.11 (1970), March 27, p. 48–51, 54–57, p. 55. This refers to a study of 84 representative card operations, prepared by Richard N. Salle and Constantine Danellis for the Charge Account Bankers Association in 1969.

47 Stearns, *Electronic Value Exchange*, p. 67.

48 O'Neil, "A Little Gift from Your Friendly Banker," p. 50.

49 Evans and Schmalensee, *Paying with Plastic*, p. 57.

50 Tony D. Sampson and Jussi Parikka, "Learning from Network Dysfunctionality. Accidents, Enterprise, and Small Worlds of Infection," *A Companion to New Media Dynamics*, ed. John Hartley, Jean Burgess and Axel Bruns (Oxford: Blackwell, 2013), p. 450–460.

to establish the credit card signs in the public mind, therefore promoting new forms of making oneself socially and economically account-able by using new ways of payment. A whole media ecology of newspapers, billboards, sports stadiums, radio and TV ads was getting mobilized to install signs of trust in almost every wallet, on almost every merchant's door.[51] Lengthening the new chains of transaction into every neighborhood and every shopping center relied heavily on mass media coverage.

Picture this as an intermedial frenzy that is quite literally targeting the suburban middle class: by mailing ever more unsolicited cards, by repeating the taglines all over again, by showcasing scenes of easy payment. It was a broadcasting situation full of individualized "to whom it may concern" messages, being delivered with plastic credit attached. A highly popular LIFE magazine article written by Paul O'Neil, published in the March 27 issue of 1970, sums up both the social troubles created by the mass mailings, and the skeptical voices on the unregulated circulation of the cards. LIFE's front cover pictured the traveling salesman "up in the air," mobilized by credit card wings. The cover served as a reminder of the elite Travel & Entertainment cards, not yet reflecting the reality that the suburban, well-educated WASP families with kids were the main users of credit cards[52] (Fig. 1).

O'Neil's article rather satirically retold the stories of banking competition and its discontents:

> In Chicago, during 1966, banks fought each other like jackals to get their cards into the hands of the public only to discover that brigades of thieves, conmen and deadbeats were galloping through stores with them and running up disastrous sums in fraudulent or uncollectable debt. [...] Credit cards, and particularly bank cards, have inspired new and enduring types of white-collar crime, have attracted the beady attention of the Mafia, and have revealed a fascinating capacity of dishonesty in employees of the postal system, who steal them from the mails and sell them for prices up to $50.[53]

Apart from these "network accidents" – which did not come as a surprise to the banks themselves – the American public already regarded computing accounting and credit data as a key technology. It was deemed necessary in handling the substantial

—

51 Stearns, *Electronic Value Exchange*, p. 207–208.

52 Mandell, *Credit Card Use in the United States*, p. 13–18.

53 O'Neil, "A Little Gift from Your Friendly Banker," p. 48–49.

Fig. 1: LIFE magazine, front cover. March 27, 1970.

paperwork involved in transactions.[54] Both BankAmericard and Master Charge already had set up regional computing centers that, given the amount of credit card fraud, were especially important for validating transactions above $50. This would typically require the merchant to call a center, reading out the customer's card number, which the telephone girl would enter into the computer terminal to crosscheck on the customer's card account.[55] This kind of everyday proof of a person's economic accountability is what occurs to O'Neil as a likely future of the medium's automation: "The bank card is an imperfect medium of exchange at the moment because nobody can be absolutely certain that it, unlike the impersonal $20 bill, is always backed by real buying power."[56]

Extending chains of financial operation to an electronic domain creates a tremendous necessity for re-mediated trust and new identification practices; certification,[57] validation, standardization, legal security, and costly public relations are needed to establish the new neighborhood technologies of credit with mobile users and shifting topologies of cooperation. Interestingly enough, the caption of a cartoon created by illustrator John Huehnergarth (Fig. 2) sums up the situation quite accurately: "In a rush to 'get their

—

54 Compare. Stearns, *Electronic Value Exchange*, p. 30–39 for "A Typical Transaction in 1968".

55 O'Neil, "A Little Gift from Your Friendly Banker," p. 56.

56 O'Neil, "A Little Gift from Your Friendly Banker," p. 56.

57 Compare Lawrence Busch, *Standards. Recipes for Reality* (Cambridge, Mass.; London: MIT Press, 2011), chap. 4.

A Little Gift from Ycur Friendly Banker

The mails bring credit cards to millions, opening new vistas for crime, chaos and comedy

by PAUL O'NEIL

Illustrated by JOHN HUEHNERGARTH

Fig. 2: LIFE magazine, March 27, 1970, p. 48–49.

plastic in the air,' banks randomly fired off credit cards. Computers – key to controlling them – are still trying to catch up."[58]

Catching up on the developments was also highly necessary for the government itself. While bank law largely remained a federal issue (leading to different amounts of interest in every state) the mass mailing campaigns got officially banned by the Federal Trade Commission in April 1970.[59] This was not only a measure that was directed at the obvious fraudulent practices. Even more so, it addressed "the unwillingness and/or inability of the banks to protect their customers from the effects of their mistakes" which had at least been willfully tolerated within overall card business strategies.[60] While numerous lawsuits were fought between banks and customers, and between the big co-opetitive networks Master Charge and Bank Americard themselves, the marketing machinery continued to launch campaign after campaign. LIFE magazine's entire volume of 1970 was full of them, too.

But how could we understand these formative years on a more theoretical level? Along with Harold Garfinkel it can be said that a new kind of sociality is getting

—

58 O'Neil, "A Little Gift from Your Friendly Banker," p. 49.

59 Evans and Schmalensee, *Paying with Plastic*, p. 73.

60 Mandell, *The Credit Card Industry*, p. 53.

invented that relies on the credit card as a new medium of account-ability in shifting mobile neighborhoods. Being part of the paper-based and electronic bureaucratic process includes a person's, a family's or a corporation's "investment in forms," as Laurent Thévenot has formulated this kind of social media interaction – in which making yourself accountable is inevitable – so aptly.[61]

Along with Michel Callon, we might understand the configuration of new chains of payment operations as one way to make a heterogeneous techno-economical network (a) convergent and (b) irreversible in its translations.[62] As Callon has shown, this is a significant element of networking practices and, as I would like to add, of the infrastructural platforms they rely on. Convergence is what actors strive for to make heterogeneous cooperation work more smoothly. Irreversibility is what actors need to normalize operations, defining standards, certification practices and common modes of exchange. This is less a static way of "blackboxing" – to which it can lead nonetheless. It is rather a way to deal with fluctuating and dynamic network properties that the actors themselves create while charging, paying, borrowing, trusting and distrusting. In the case of the American credit card, the techno-economical networks rely on a significant organizational form of convergence and irreversibility: the two-sided co-opetitive platform market. What kind of neighborhood technology is it, how did it evolve and what does it mean for a media history of the credit card?

4. BUILDING CO-OPETITIVE PLATFORMS FOR NETWORKING

The year 1966 certainly was *the* annus mirabilis for the emergence of the credit card's infrastructures. American Express, Carte Blanche and Diners Club were well established by then, leading the field of Travel & Entertainment Cards as "go-it-alones". Few of the other experimental card systems had gained momentum, although Bank of America had managed to establish its BankAmericard program in California, earning its first operating profit in 1961. The extension of operations to the national scale did not start before 1966, when Bank of America offered to license its card to selected banks across the

---

61 Laurent Thévenot, "Rules and Implements: Investment in Forms," *Social Science Information* 23.1 (1984), p. 1–15.

62 See also Michel Callon, "Techno-Economic Networks and Irreversibility," *A Sociology of Monsters? Essays on Power, Technology and Domination*. Ed. John Law (London; New York: Routledge, 1991), p. 132–161.

country. This was in fact a brilliant coup since regulations had prohibited the bank to open branches in other states.[63]

This, in return, led to quickly emerging local associations of banks that did not become part of the franchise. As opposed to American Express and Diners Club who attempted to put up franchises of their own, the local associations in Chicago, Michigan, New York City, Buffalo, Pittsburgh, Milwaukee, Seattle and Phoenix formed an "Interbank Card Association." The Interbank members very soon cooperated with the Californian "Western States Bankcard Association" that had created the Master Charge brand in 1966. Both networks joined in 1967.[64] In 1968, the results of this rush in organizational changes became visible: "there were now two competing national networks of banks: the BankAmericard franchise system, and the Interbank cooperative system."[65] The Interbank system – recognizable by the black "i" on the Master Charge cards [Fig. 1] – attracted more members. In response, Bank of America agreed to reorganize the BankAmericard structure into another membership-owned corporation in 1970, calling it National Bank Americard Inc. (NBI).

Both the Interbank Association and NBI quickly proved to be highly paradoxical infrastructures of banking. Clearly, the member banks were competing with each other concerning interest rates, fees and services, bombarding the upper und middle class with mass mailed unsolicited cards. Yet they mostly agreed to cooperate by setting standards for interoperability in analogue and digital accounting, contracting merchants and common advertising early on. This small cooperative core is where Callon's convergence and irreversibility trajectories play a highly important role. Mixing cooperation and competition leads to what economists call "co-opetition" – a rather astonishing way of enrolling both merchants and customers to mobilize payment and credit on a worldwide scale. Still, competition would also be upheld in the long run by the duality between Master Charge (MasterCard) and BankAmericard (Visa), with American Express and Discover as strong single corporations in a "four hub ecosystem".[66]

I cannot get into the full economic and historical details here, but the phenomenon of "co-opetition" is intricately connected to the building of platforms that serve as a common basis for "cooperation without consensus," to borrow from Susan Leigh

---

63 Stearns, *Electronic Value Exchange*, p. 26.

64 Evans and Schmalensee, *Paying with Plastic*, p. 59–64.

65 Evans and Schmalensee, *Paying with Plastic*, p. 64.

66 Evans and Schmalensee, *Paying with Plastic*, p. 178.

Star and James Griesemer's famous article on heterogeneous cooperation.[67] The credit card platforms have been understood as forerunners of comparable business models that need to assemble two or more distinct groups of customers, being likely to be very heterogeneous.[68] Running such a platform must be something that promises economic surplus for the intermediary actors, a third actor or parasite, if you will. Losing money on transaction costs might be acceptable, if the intermediary profits from the overall relationship.[69] Turning platforms into a neighborhood for accountable actions has become one peculiar and highly successful style of informational capitalism. "Platform politics"[70] thus clearly predate the era of the Personal Computer and the World Wide Web, and they are highly likely to be indicators for the intertwined histories of socio-economic and media practices. Getting to terms with social media might necessarily involve attention to bureaucratic characteristics of all kinds of platforms. This even exceeds the openly techno-economical variants and may as well extend to a Science and Technology Studies based "platform sociology," as proposed by Peter Keating and Alberto Cambrosio for the case of biomedical platforms:

"Insofar as they embody regulations and conventions of equivalence, exchange, and circulation, platforms are not simply one among many forms of coordination that include networks or, rather, they *account* for the generation of networks or, at the very least, they are a condition of possibility for the very existence and transformation of networks. The intermediaries that stabilize networks are produced and reproduced on the platform. Platforms supply networks with conventions, generate novel entities, and entrench them in clinical routines".[71] Put shortly, they provide means of convergence and architectures of irreversibility (Callon), while being themselves battlegrounds of standardization and certification of operational chains, and overall "maintenance and repair".[72]

—

67 Susan Leigh Star and James R. Griesemer, "Institutional Ecology, 'Translations' and Boundary Objects: Amateurs and Professionals in Berkeley's Museum of Vertebrate Zoology, 1907–39," *Social Studies of Science* 19.3 (1989), p. 387–420.

68 Rochet and Tirole, "Platform Competition in Two-sided Markets".

69 Evans and Schmalensee, *Paying with Plastic*, p. 244.

70 Tarleton Gillespie, "The politics of 'platforms'," *New Media & Society* 12.3 (2010), p. 347–364; Joss Hands: "Introduction: Politics, Power and 'Platformativity'," *Culture Machine*, *Issue "Platform Politics"* 14 (2013). URL: http://www.culturemachine.net/index.php/cm/article/view/504/519.

71 Peter Keating and Alberto Cambrosio, *Biomedical Platforms. Realigning the Normal and the Pathological in Late-Twentieth-Century Medicine* (Cambridge, Mass.; London: MIT Press, 2003), p. 324. Emphasis added.

72 Stephen Graham and Michael Thrift, "Out of Order. Understanding Repair and Maintenance," *Theory, Culture & Society* 24.3 (2007), p. 1–25.

Yet they are also foremost a neighborhood "technology of accountability," being "active, generative, and opaque" at the same time.[73] Platforms provide a veritable playing field for both the irreducible agency *and* the regulation of their assembled media practices. And they are the bureaucratic writing systems that attempt to stabilize social media interaction, in which "stability" of course is always an unlikely event of orderliness. Or, as Harold Garfinkel would put it: There are usually good organizational reasons for bad actuarial records of "what was going on while it was going on" on a given platform.[74]

5. DIGITAL MOMENTUM: TOWARDS A MEDIA HISTORY OF THE CREDIT CARD

The co-opetitive networks quickly called out for a standardization of their business, but did not manage to develop interoperability by themselves. Rather, they contacted the American National Standards Institute (ANSI) in 1968, which managed to publish the first standards in 1971. While the issues at stake here might seem trivial, they were not taken lightly by the involved actors: card dimension, location of the signature panel, font and format of the embossed characters, an account numbering system for interchange, plus the magnetic strip (whose track features caused a huge controversy in the industry) needed an impartial unification.[75] And so did a multitude of other technical innovations in computing and media: optical character recognition of signatures (OCR), automated telling machines (ATM), chips-on-a-card,[76] early special-purpose digital networks,[77] and security technologies like holograms.[78] The card business certainly has pushed the history of digital media forward. Yet, since all of these inventions further mobilize social practices of crediting and indebting, they cannot be seen only as another grand example of sociotechnical "systems building" (Thomas P. Hughes) designed for multi-sided platform markets.

—

73 Keating and Cambrosio, *Biomedical Platforms*, p. 326.

74 Garfinkel, "Good organizational reasons for 'bad' clinical records".

75 Almarin Philipps, "The Role of Standardization in Shared Bank Systems," *Product Standardization and Competitive Strategy* (Amsterdam: Elsevier, 1987), p. 273.

76 Mandell, *The Credit Card Industry*, p. 143–152.

77 See also Stearns, *Electronic Value Exchange*, p. 71–108, 135–156.

78 Compare Jens Schröter, "The Age of Non-Reproducibility," *Film and Media Studies (Acta Univ. Sapientiae)* 5 (2012), p. 7–20, p. 13.

It might also not be enough to refer to American neighborhood transformations only, as Lewis Mandell did in his seminal history: "The credit card has evolved to meet the needs of our mobile, affluent society. Credit has always been available from local merchants, but as mobile customers began dealing with merchants over a larger geographic area, the personal trust that had existed between the merchant and his regular customers had to be replaced with the trust of a well-known third-party guarantor."[79] Issued by these third parties, credit cards serve as "immutable mobiles" – be they trusted or not – for the global extension of payment platforms and their personal accounts.[80]

Within a world history of credit, the first ANSI standards of 1971 would certainly overlap with the political-economical downturn of the static Bretton Woods system, and the switch back to freely fluctuating exchange rates of currencies. A media history of credit might also account for a different non-hegemonic theory of money. Credit and debt would be at its core, embodying the *fait accompli* of most systems of human exchange of signs, commodities, and their agency in social cooperation.[81] The gifts of the credit card certainly are a case for the innocent anthropologist. But let's charge that to another account.

—

79 Mandell, *The Credit Card Industry*, p. 154.

80 Compare Bruno Latour, "Visualisation and Cognition: Drawing Things Together," *Representation in Scientific Practice*. Ed. Michael Lynch and Steve Woolgar (Cambridge, Mass.; London, England: MIT Press, 1990), p. 19–68.

81 Graeber, *Debt. The First 5,000 years*, chap. 3. Graeber's reading of Keynes is cutting it to the point: "Money is credit," ibid., p. 54.

IV. NEIGHBORHOOD ACTIVITIES

SHINTARO MIYAZAKI

NEIGHBORHOOD SOUNDING

AN ARCHAEOLOGY OF DYNAMIC MEDIA NETWORKS 1960–1980 | 2010

ABSTRACT

According to the French collective of mathematicians which published under the pseudonym Nicolas Bourbaki, neighborhood as a term used in topology is defined as an expression for "sufficiently near" or "as near as we please."[1] In the monograph *Micromotives and Macrobehavior* by sociologist Thomas C. Schelling the concept of neighborhood plays an important role for analyzing the dynamics of separation and segregation of ethnic groups.[2] And the jargon of telecommunications uses the term *neighbor* inter alia for describing the closest switching computers in *packet switching* – a technical term for the media technology of early Internet.

This contribution is a media archaeological inquiry into the past of dynamic media networks such as the Internet and a close reading of some early historical publications describing early computer networks. Concentrating on the programming of routing procedures it outlines *neighborhood sounding* as a key moment of packet switching, an early form of distributed and dynamic networks, where each agent or node was defined by the lively exchanges of its neighboring nodes. It finally discusses whether the assemblage[3] or agencement of *ARPAnet* can be regarded as a complex system,[4] showing emergent behavior or not and whether packet switching can be perceived as an early implementation of principles embodied in more advanced neighborhood technologies of the 21st century.

—

1 Nicolas Bourbaki, *Elements of Mathematics. General Topology. Chapters 1–4 [Éléments de Mathématique. Topologie Générale, Paris: Diffusion C.C.L.S. 1971]* (Berlin: Springer, 1989), p. 19.

2 Thomas C. Schelling, *Micromotives and Macrobehavior. With a New Preface and the Nobel Lecture*, 1st Ed. 1978 (New York/London: W. W. Norton & Company, 2006), p. 145.

3 Manuel DeLanda, *New Philosophy of Society: Assemblage Theory and Social Complexity* (London/New York: Continuum, 2006).

4 Melanie Mitchell, *Complexity. A Guided Tour* (Oxford/New York: Oxford University Press, 2009).

1. NEIGHBORHOODS, MEDIA AND NETWORKS

By analyzing the quantitative dynamics of residential segregation in the early 1970s, Thomas C. Schelling defined neighborhood as a spatial arrangement of households, where each neighbor is defined in reference to their own location.[5] According to Schelling, segregation is a dynamic process where "small incentives, almost imperceptible differentials, can lead to strikingly polarized results."[6] The principle that small changes in the micro-behavior of a neighborhood relate to the macro-behavior of the whole system is also relevant for dynamic media networks. Paul Baran (1926–2011), who worked at the department for Computer Science and Mathematics at RAND Corporation since 1959, wrote and conducted a study on distributed communication. It was presented in 1961 and finally published in 1964.[7] It is notable in this context that also Schelling's research was funded by the RAND Corporation, but ten years later than Baran's study.

Baran conceptualized and simulated a network of computers with locally implemented neighborhood dependent switching rules, which he called a distributed adaptive message-block network.[8] The primary aim of this project was to inquire into the technical implications of building a communication network which could survive a major hostile military attack.[9] The layout of Baran's network was sketched out as a mixture of telephone lines – that is, to be more precise, of pulse-code-modulated multiplexing systems developed by the Bell Labs –, and of microwave stations and satellites networks.[10]

2. SOUNDING ITS NEIGHBORHOOD

Distributed networks are effectuated by the specific interplay of two instances: The node and the packet. The packet is the smallest unit of transmission (in terms of time) and the node is the smallest unit of the network (in terms of space). A packet, as conceptual-

—

5 Thomas C. Schelling, "Dynamic Models of Segregation," *Journal of Mathematical Sociology* Vol. 1 (Gordon and Breach Science, 1971), p. 147–148.

6 Schelling, "Dynamic Models of Segregation," p. 146.

7 Paul Baran, *On Distributed Communications: I. Introduction to Distributed Communication Networks (Memorandum, RM-3420-PR, August 1964)* (Santa Monica, CA: The RAND Corporation, 1964): p. iii.

8 Baran, *On Distributed Communication*: p. iii.

9 Baran, *On Distributed Communication*, p. 1.

10 Baran, *On Distributed Communication*, p. 18–19.

ized by Baran, has a standardized and fixed sequence of sections containing data such as (1) start of message, (2) address, (3) sender, (4) handover number, (5) content and (6) end of message. These sequences thus were quite similar to the structure of machine code instructions used in early mainframe computing, albeit slightly more complicated.[11] A constitutive condition for the working of such a network is the timing, flow control or the algorhythmics[12] of the procedure how the small data packets will get forwarded by the nodes and arrive at their point of destination. In this context, Baran proposed an algorithm or method called *store-forward*, based on the metaphor of a postman:

> "A postman sits at each switching node. Messages arrive simultaneously from all links. The postman records bulletins describing the traffic loading status for each of the outgoing links. With proper status information, the postman is able to determine the best direction to send out any letters. [...] [T]he postman can infer 'best' paths to transmit mail by merely looking at the cancellation time or equivalent hand over number tag. [...] Each letter carries an implicit indication of the length of transmission path."[13]

While the metaphor of an intelligent and skilled ultra-fast human postman is easy to understand, I would like to argue that another description gets significantly closer to the non-human technological workings of distributed networks: It is the model of *neighborhood sounding*, namely of a machine sending signals to and listening to signals from its neighborhood according to a fixed, algorithmically defined ruleset. Etymologically, sounding in its conventional meaning derives from the French term *sonde* (in English: *probe*) and refers to the mechanism of probing the environment – most often distances – by sending out an object and observing the physical responses it caused during its trajectory. Sometimes the object would get reflected and returns back, sometimes it is tied to a string or thread that will act as a medium and sometimes not an object, but a signal is sent out. The last practice is connected to a more unconventional meaning of sounding, which pertains to sending out a sound, not a probe, thus sounding derived from sound, not *sonde* and therefore the performative act of playing a signal and listening for its responses. The medium of sounding changed from being object-based

---

11 Baran, *On Distributed Communication*, p. 22.

12 Shintaro Miyazaki, "Algorhythmics: Understanding Micro-Temporality in Computational Cultures," *Computational Cultures. A Journal of Software Studies* 2 (2012), http://computationalculture.net/article/algorhythmics-understanding-micro-temporality-in-computational-cultures (January 2014).

13 Baran, *On Distributed Communication*, p. 25.

to operating signal-based. An early manifestation of this shift, for instance, has been the measurement of distances underwater.[14] On this basis *neighborhood sounding* describes a software-programmed mode of *sounding, that is, measuring with signals,* within a network of multiple communication channels such as computer networks. Furthermore, *neighborhood sounding* refers to a general principle of listening, sensing, reacting and acting implemented in animal echolocation as well as in human sonar operator skills. Physicians practiced percussive diagnosis by sounding out the body[15] before the dawn of medical ultrasonography and other imaging technology. Other sounding practices were applied in classical archaeology for inquiring into the conditions of the ground to excavate. And in a surprisingly similar way, the delays and transmission times in telegraphy networks were measured by sending out electrical impulses, thus sounding out the network. Later, comparable techniques were applied to telephone networks. While the informed reader might remark that acoustic and electromagnetic signals are not the same, he or she might remember that these can easily get transduced[16] into each other as acoustic-couplers the first modems for digital networks showed from early on. Acoustics is one of the first sciences where the practice of modeling general principles by equivalence circuits got established.[17]

3. SIGNALS AND RHYTHMS OF ARPANET

Activated in the late 1960s, the ARPAnet was the first "large-scale demonstration of the feasibility of packet-switching".[18] ARPAnet's major difference to later forms of distributed networks – which were based on the TCP/IP-Protocol after its establishment in the early 1980s – consisted of its *Interface Message Processors* (IMPs). These mediated the handling of

---

14 John Shiga, "Sonar: Empire, Media, and the Politics of Underwater Sound," *Canadian Journal of Communication* 38, no. 3 (September 2013), p. 357–77; Willem Hackmann, *Seek and Strike. Sonar, Anti-submarine Warfare and the Royal Navy 1914–54* (London: Stationery Office Books, 1984).

15 Axel Volmar, "Listening to the Body Electric. Electrophysiology and the Telephone in the Late 19th Century," *The Virtual Laboratory* (Berlin: Max-Planck-Institute for the History of Science, 2010). Http://vlp.mpiwg-berlin.mpg.de/essays/data/art76.

16 Adrian Mackenzie, *Transductions. Bodies and Machines at Speed* (London/New York: Continuum, 2002).

17 Roland Wittje, "The Electrical Imagination: Sound Analogies, Equivalent Circuits, and the Rise of Electroacoustics, 1863–1939," *Osiris* 28, no. 1 (January 1, 2013): p. 40–63.

18 Janet Abbate, *Inventing the Internet* (Cambridge, MA: MIT Press, 1999), p. 7.

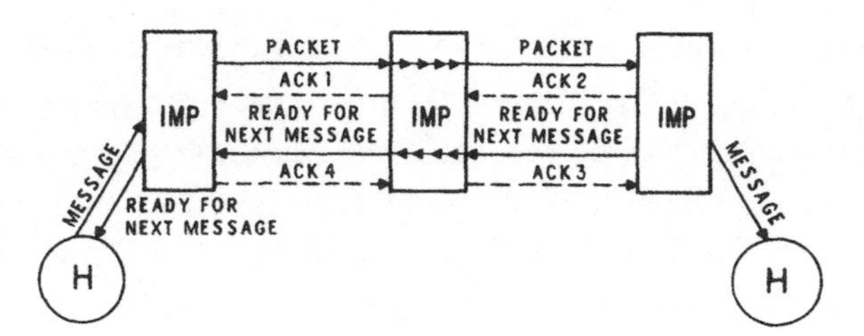

Fig. 1: RFNMs and acknowledgments, Fig. 4 in F. E. Heart, R. E. Kahn, and S. M. Ornstein, "The Interface Message Processor for the ARPA Computer Network," in *AFIPS '70 (Spring) Proceedings of the May 5–7 1970, Spring Joint Computer Conference* (New York: ACM, 1970), 554.

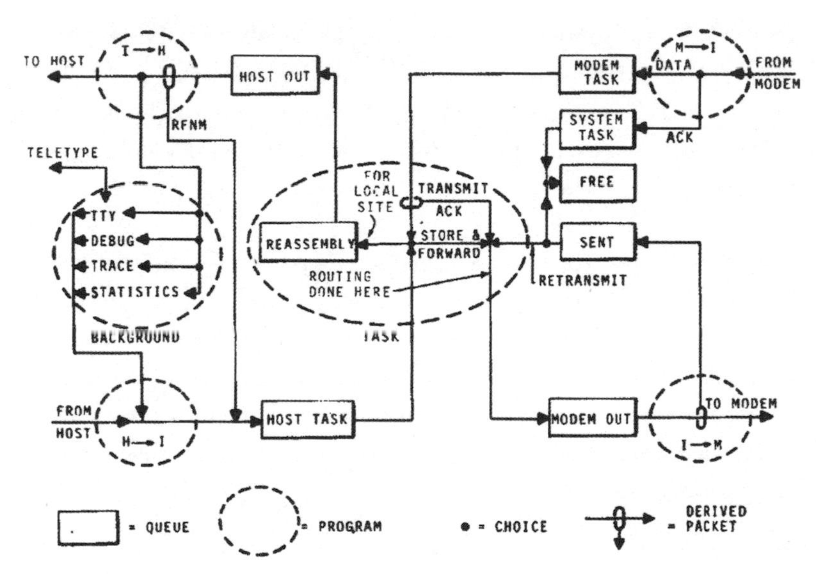

Fig. 2: Internal packet flow, Fig. 9 in ibid., 561.

the messages and packets between two host computers.[19] IMPs were small computers – such as a Honeywell DDP-516 – installed only for packet switching. As shown in Fig. 1, the timing of a packet transmission consists of multiple layers. A message of digital data gets divided into smaller packets, which are individually transmitted over the network towards their final destination. When the topology – that is, the structure – of the

—

19 F. E. Heart, R. E. Kahn and S. M. Ornstein, "The Interface Message Processor for the ARPA Computer Network," *AFIPS '70 (Spring) Proceedings of the May 5–7 1970, Spring Joint Computer Conference* (New York: ACM, 1970): p. 551–567.

network changes due to a failure or an interruption somewhere on their way, the packets can take alternative and different routes. After arriving at the final destination, they are assembled into the right order again. Infrastructural signals for ensuring an error-free transmission such as the *Ready for Next Message* (RFNM) and the acknowledgment (ACK) signals were most important for enabling the *sounding*. Fig. 2 shows an abstract diagram of the data packet flow inside such an IMP, which operates as a medium between a host computer and the modem (acoustic coupler). The packet processing is organized in sub-routines and programs notably as well the background functions such as trace, debug and statistics, which would become important for understanding unexpected breakdowns and failures.

The adaptive routing algorithm implemented in the IMP defined the procedure of packet switching. It tries to estimate the path with the shortest transit time to the final destination. The main principle of such an algorithm is that it uses data tables about the condition of the neighboring network, its connectivity, or signs of traffic congestions. This data gets updated every half second.

> "Each IMP estimates the delay it expects a packet to encounter in reaching every possible destination over each of its output lines. It selects the minimum delay estimate for each destination and periodically (about twice a second) passes these estimates to its immediate neighbors. Each IMP then constructs its own routing table by combining its neighbors' estimates with its own estimates of the delay to that neighbor. The estimated delay to each neighbor is based upon both queue length and the recent performance of the connecting communication circuit."[20]

The timing of each operation in each IMP is consequently highly relevant.[21] Even in the absence of packets to transmit an IMP sends its neighbors a "hello" packet and expects a "I heard you" packet within a period of 0.5 seconds. A dead line is defined by a sustained absence of about 2.5 seconds of such messages.[22]

—

20 Heart et al., "The Interface Message Processor for the ARPA Computer Network," p. 555.
21 J.M. McQuillan, G. Falk and I. Richer, "A Review of the Development and Performance of the ARPANET Routing Algorithm," IEEE *Transactions on Communications* 26, no. 12 (December 1978).
22 Heart et al., "The Interface Message Processor for the ARPA Computer Network," p. 555.

4. BREAKDOWNS AND COMPLEX SYSTEMS

The rigidly programmed neighborhood relations and operativity of one IMP, or, in other words, its micro-behavior, affects the overall macro-behavior of the network and its communication processes: "A local failure can have global consequences."[23] Consequently, the early ARPAnet is a complex system as defined by Schelling. It "exhibits nontrivial emergent and self-organizing behaviors."[24] This especially becomes apparent in case of unexpected breakdowns and disturbances. While emergence is usually reserved to living organisms and other ecological assemblages, a reading of the early descriptions of the ARPAnet pioneers reveals that this attribution might as well apply to the nonorganic media ecologies they were confronted with on a daily basis.

Right from the start, Leonard Kleinrock of University of California was one of the researchers who had to deal with the complexity of distributed networking. In a *Proceedings of the* IEEE paper, published in 1978 in a special issue on packet switching, he noted: "Not only was the demand process bursty, it was also highly unpredictable in the sense that the instants when the demands arose and the duration for the demands were unknown ahead of time."[25] Kleinrock recognized that the probabilistic complexities of distributed networks are "extremely difficult" and that an effective "flow control" within the network was one of the important needs, which first was underestimated and ignored.[26] Such flow control mechanism could have prevented many deadlock phenomena such as "reassembly lockup," "store and forward deadlock," "Christmas lockup" and "piggyhack lockup."[27]

Such system deadlocks had been discussed, researched and prevented already in the era of time-sharing during the 1960s,[28] but in case of the ARPAnet the media-ecological environment was less optimal than before. The new network worked with less bandwidth and was geographically distributed over the whole North-American continent, where as the earlier time-sharing, multi-user and multi-programming networks such

---

23 McQuillan et al.,"A Review of the Development and Performance of the ARPANET Routing Algorithm," IEEE *Transactions on Communications*, Vol. 26 , Nr. 12, Dec., p. 1805.

24 Mitchell, *Complexity. A Guided Tour*, p. 13.

25 Leonard Kleinrock, "Principles and Lessons in Packet Communications," *Proceedings of the* IEEE. *Special Issue on Packet Communications*, Vol. 66, No. 11, Nov., ed. Robert E. Kahn (New York: IEEE, 1978), p. 1321.

26 Kleinrock, "Principles and Lessons in Packet Communications," p. 1322.

27 Kleinrock, "Principles and Lessons in Packet Communications," p. 1324.

28 E. G. Coffman, M. Elphick and A. Shoshani, "System Deadlocks," *ACM Comput. Surv.* 3, no. 2 (June 1971), p. 67–78.

as the famous *Compatible Time-Sharing System* developed at Massachusetts Institute of Technology or *Programmed Logic for Automatic Teaching Operations* a system designed and built by the University of Illinois used more bandwidth and less concrete physical space, thus were in general more controllable.[29] As an effect, Bolt, Beranek and Newman, an East coast-based company which produced the IMPs, implemented measurement tools – sounding tools – to detect the network behavior from the outset.[30] Without these observation media the control, research and optimization of the network would never have been possible in the first place. The ARPAnet was an *experimental system* much in the sense of German historian of science Hans-Jörg Rheinberger.[31] It was not a static system, but was adjusted continuously and optimized according to knowledge gained from measuring and observing unpredicted failures and not by sophisticated prediction. Furthermore the network was not only working with adaptation on the algorithmic level in the IMPs, but also on the level of the many man-made adaptations these algorithmic control mechanisms would need to make the system more stable.[32] Kleinrock described this recursive entanglement as "philosophical":

> "It is ironic that flow control procedures by their very nature present *constraints* on the flow of data (e.g., the requirement for proper sequencing), and if the situation ever arises whereby the constraint cannot be met, then, by definition, the flow will stop, and we will have a deadlock! This is the philosophical reason why flow control procedures have a natural tendency to introduce deadlocks."[33]

This epistemological reasoning is extendable to digitized computer-based communication in general. The introduction of complicated *algorhythmics* to control the data flow in complex media networks require careful programming and measurement methods. Otherwise, the system will stop processing. Adaptation and revision of the important concepts and the measurement procedures due to unpredictable failures is thus a core

—

29 Kleinrock, "Principles and Lessons in Packet Communications," p. 1322.

30 G. Cole, "Performance Measurements on the ARPA Computer Network," IEEE *Transactions on Communications* 20, No. 3 (June 1972), p. 630 – 636.

31 Hans-Jörg Rheinberger, *Toward a History of Epistemic Things: Synthesizing Proteins in the Test Tube* (Stanford, CA: Stanford University Press, 1997).

32 See for an overview of some of these adaptation processes, McQuillan et al., "A Review of the Development and Performance of the ARPANET Routing Algorithm".

33 Kleinrock, "Principles and Lessons in Packet Communications," p. 1324.

agent in the historical transformation and optimization processes of such *algorhythmic networks* as the ARPAnet.

Thinking about the role of non-human agency in dynamic media networks the philosopher of complex assemblages and former software programmer Manuel DeLanda remarked: "[W]hile computers were originally seen as the medium for getting men out of the loop, network decentralization introduces a new kind of independent will, the independent software objects (or demons), which might prove as difficult to enslave as their human counterparts."[34]

With the network decentralization introduced by the ARPAnet failures and breakdowns such as those deadlocks described above provoke the interpretation that they exhibit emergent and self-organizing behaviors. It seems that with the dawn of distributed networking a new epistemic metaphor began to evolve. It was the model of media networks as complex non-human media ecologies. This shift towards ecologies provokes new theoretical concepts for their historical inquiry.

5. GOING BEYOND ARCHAEOLOGIES

Neighborhood sounding implies not only an intensified focus on the timing, delays and rhythms of dynamic media networks, but calls as well for a heterogeneous, multi-scalar analysis of their immanent processes, interdependencies and topological structures. This means that the metaphor of an *archaeologist* who works on the material conditions of past cultures influenced by Michel Foucault[35] and advocated by recent media theorists such as Wolfgang Ernst, Siegfried Zielinski, Erkki Huhtamo or Jussi Parikka[36] needs an extension towards the metaphor of an *ecologist* who reaches present complex systems using specific media-based observation methods: "Ecologists focus rather more on dynamic systems in which any or one part is always multiply connected, acting by

—

34 Manuel DeLanda, *War in the Age of Intelligent Machines*, First Edition 1991 (New York: Zone Books, 2003), p. 121.

35 Michel Foucault, *The Archaeology of Knowledge and The Discourse on Language* (translated from French by A. M. Sheridan Smith) (New York: Pantheon Books, 1972), p. 135–140.

36 Erkki Huhtamo and Jussi Parikka, eds., "Introduction: An Archaeology of Media Archaeology," in *Media Archaeology. Approaches, Applications, and Implications* (Berkeley, CA: University of California Press, 2011), p. 1–25.

virtue of those connections, and always variable, such that it can be regarded as a pattern rather than simply as an object."[37]

Emphasizing the rhythmic and dynamic aspects of both past and present complex networks might help to understand their complexity a little more than just looking at their topological or *objective* properties. Under such assumptions, *neighborhood sounding* implies methods for understanding media in quite an active and invasive way. It requires an engagement with the neighborhood in question. *Neighborhood sounding* is not only a principle that is constitutive and pivotal for the inner working of distributed and complex networks. It is also an important methodology in order to inquire them. While trying to understand the certain dynamics of present communication networks and ecologies such as Twitter, Facebook, Instagram, Flickr, processes in the World Wide Web in general or of the past such as the ARPAnet the critical researcher needs to select different nodes of such networks both real or simulated, conduct an analysis of the incoming and outgoing data signals and inquire these rhythms and their timing under his or her specific research question. In doing that he or she might need to go beyond the own disciplinary boundaries of humanities or sociology, acclimate oneself to emerging and long-forgotten neighborhoods, and build his or her own information-aesthetical devices and systems of exploration, experiments and sounding by incorporating current methods of network analysis and measurements both in the study of software and, if possible, even of hardware.

---

37 Matthew Fuller, *Media Ecologies. Materialist Energies in Art and Technoculture* (Cambridge, MA: MIT Press, 2005), p. 4.

CAROLIN WIEDEMANN

DIGITAL SWARMING AND AFFECTIVE INFRASTRUCTURES

A NEW MATERIALIST APPROACH TO 4CHAN

ABSTRACT

In this essay on the notion of the online swarm I suggest that the euphoric appreciation of the democratic potential of the Internet still disregards the complex entanglements of affects and infrastructures within processes of constitution in biopolitical societies of control. In order to approach these entanglements, I investigate the online swarm with a neo-materialist perspective and consider a specific Internet infrastructure, 4chan, the board on which Anonymous first emerged, as an agent within the swarm formation. After discussing the methodological challenges implied by this perspective, I will approach the infrastructural mediators by analyzing their affective dimension. I conclude by posing a question: If we look at the phenomenon from this theoretical perspective, can we then still speak of emancipation and solidarity?

INTRODUCTION

From Howard Rheingold's *Smart Mobs*[1] in 2003 to Felix Stalder's concept of *Digital Solidarity*[2] in 2013, the Internet has inspired and still inspires the dream of new sorts of collectivity, of a potentially free and open space of information and communication that would emancipate and unite the people. These discourses often employ the swarm metaphor, the "ephemeral and apparently 'grass-roots democratic' conception of collectivity"[3] that suggests emergent cooperation or solidarity and is therefore also used to point to new emancipatory politics.

The notion of a solidary swarm is especially interesting with regard to present forms of governmentality: On the one hand, it stands for subversive moments within societ-

---

1 Howard Rheingold, *Smart Mobs. The Next Social Revolution* (New York: Basic Books, 2003).

2 Felix Stalder, *Digital Solidarity* (Lüneburg: Post-Media Lab, 2013).

3 Sebastian Vehlken, "Zootechnologies: Swarming as a Cultural Technique," *Theory, Culture & Society* 30(6) (2013), p. 110–131, here: p. 112.

ies of control or biopolitical capitalism, as it can point to new forms of sociality that overcome logics of neoliberal competition.[4] On the other, inspiring swarm behavior is one way of governing in neoliberal capitalism. As the notion of the online swarm and, more generally, of constituting collectivity on the Internet has these political dimensions, it is worth approaching it again from the *neighborhood technologies* point of view. This project aims, among other things, to develop a more differentiated and technologically informed notion of neighborhood concepts, including *swarm intelligence*, a notion that takes into account the role of media technologies. The material turn partly has a similar focus: Against a cultural studies or sociological focus on semantics and meaning and the concomitant disinterest in technology as well as affects, new materialist theories ask for transdisciplinary approaches that consider agency as distributed across all things – human and nonhuman.[5] Referring to this perspective, I begin by taking into account the role of affect circulation that constitutes the online swarm. In order to illustrate these constitutive forces, I will focus on one prominent phenomenon, *Anonymous*. I will outline a special case of swarm behavior, the creation of the LOLcats, and then, still employing a new materialist perspective I will focus on the infrastructure of the site 4chan, where these swarms are still emerging. I do not want to reduce the various (inter-)disciplinary connections and diverse undertakings that are subsumed under the label of the *material turn* to a set of shared qualities. I will briefly discuss the methodological challenges that they imply and highlight some of the differences in the approaches in order to explain which approach informs my work here. The perspective focusing on the relations of online architectures and their affective potential will then allow to grasp the processes of mediation within which circular reactions create a new collectivity such as the online swarm. In the end I aim to outline to what extent understanding affective infrastructures can help to grasp online swarming in its diverse manifestations and thus might help to distinguish between swarms for example as parts of marketing strategies or swarms that rather work as subversive elements within current forms of capitalism.

---

4 Carolin Wiedemann, "'Greetings from the Dark Site of the Internet' – Anonymous und die Frage nach der Subversion in Zeiten der Informatisierung," *Österreichische Zeitschrift für Soziologie* (forthcoming).

5 In recent years new materialism more and more pushes back the transcendental, humanist and thus dualist traditions that still shaped cultural theory, as Rick Dolphijn and Iris van der Tuin write. They euphorically claim that for the first time this approach, the neo-materialist, is 'getting the attention it needs, freeing itself from the Platonist, Christian, and Modernist rule under which it suffered for so long'. See: Rick Dolphijn and Iris van der Tuin: *New Materialism: Interviews and Cartographies* (Ann Arbor: Open Humanities Press, 2012), p. 94–95.

1. UTOPIAN VISIONS AND CONSCIOUS SWARMS

Utopian visions accompanied the development of the Internet from its beginning, especially in the euphorically speculative period of the 1990s, when the Internet only started to become a part of everyday life.[6] The ideas of emancipation through new forms of collaboration online were taken up at the beginning of the 21st century, when the Internet was redefined as Web 2.0. In concepts of *social media* involving participation, grassroots democracy, free access for all, and nonhierarchical communities, an instrumental understanding of the media again either reduces technology to the provision of secondary technical tools or completely dismisses it.[7] Isabell Otto and her colleagues assume that in the current media culture the notion of "networking in order to form intelligent collectives or a hierarchy-free 'power of the many'" within online swarms is still widely accepted.[8]

The instrumental understanding of media infrastructures on which these approaches often rest is tied to notions of subjectivity and sovereignty that allow for an idea of emancipation in the modernist sense. For example, Felix Stalder describes the "contemporary swarm" as "a coordinated mass of autonomous, self-conscious individuals," "a self-directed, conscious actor, not a manipulated unconscious one,"[9] as opposed to unconscious manipulated crowds. This theoretical opposition expresses a notion of sovereignty and conscience (in the Marxist sense) that is not shared by (post-)structuralist or neo-materialist approaches, some of which employ the swarm metaphor but in a completely different way. Eugene Thacker, for example, in his article on *Networks, Swarms, Multitudes*, focuses on examples of mutations in the contemporary body politic and develops a notion of the swarm that is part of a concept which exceeds given paradigms of intentional subjects.[10] Similarly, Jussi Parikka notes that "insects are the privileged case study" for technologically and politically new ways of organization where

---

6 See, for example, Howard Rheingold, *The Virtual Community. Homesteading at the Electronic Frontier* (Reading MA: Addison-Wesley, 1993) and Sherry Turkle, *Life on the Screen: Identity in the Age of the Internet* (New York: Simon & Schuster, 1995)

7 Isabell Otto und Samantha Schramm, "Media of Collective Intelligence," (Research Programme of DFG-Network), http://www.uni-konstanz.de/mki/?page—id=286 (last accessed May 23, 2014).

8 Isabell Otto and Samantha Schramm stress that this is particularly remarkable in view of a culture-historical tradition in which swarms are imagined as diffuse enemies that are uncontrollable or employed by natural sciences to describe societies.

9 Stalder, *Digital Solidarity*, p. 41.

10 Eugene Thacker, "Networks, Swarms, Multitudes," *Ctheory*. May 18, 2004, part one: http://www.ctheory.net/articles.aspx?id=422 and part two: http://www.ctheory.net/articles.aspx?id=423 (last accessed May 23, 2014).

the many preexist the one, where animal packs function without heads (without one specific reason or leader); insect swarms thus "suggest logics of life that would seem uncanny if thought from the traditional subject/object point of view".[11] Proceeding from Thacker's definition of the swarm, affects become a central element, as opposed to the stance of conscious sovereignty in the utopian views of Internet collaboration.

2. SWARMS AND CIRCULAR AFFECTION: LOLCATS

Like Stalder, Thacker describes the swarm as decentralized, self-organizing, and spontaneous.[12] According to both of them, a swarm is by definition a directional force that is without centralized control but works as a collective because it has a spontaneous purpose. But while for Stalder the central force that creates the swarm comprises individual conscious actors, in Thacker's view the spontaneous purpose of the swarm cannot be traced to any of the individuated units of the swarm, only to their circular affection.

The concept of affect, derived from Spinoza, does not imply any notion of intention or conscience or the reason and motivations of individual actors; as a prepersonal phenomenon it is opposed to *feeling* and *emotion*. I join a great number of affect scholars by moving away from concepts of feeling or emotion in the definition I employ here and instead delineate affect as the "nonlinear complexity out of which the narration of conscious states such as emotion are subtracted, but also ...'a never-to-be-autonomic remainder'".[13] This new materialist approach conceptualizes affect as belonging to the bodily sphere, as a phenomenon involving (human and nonhuman) bodies that can interrupt and irritate discursive patterns but, importantly, is not an asocial phenomenon,[14] a point that will be further elaborated. Within this Spinozian perspective, affections between human and nonhuman bodies or materialities emerge and operate beyond human perception.[15] Affect, as Eva Horn explains in her reflection

---

11 Jussi Parikka, "Politics of Swarms: Translations between Entomology and Biopolitics," *Parallax*, 14: 3 (2008), p. 112–124, here: p. 115.

12 Stalder, *Digital Solidarity*, p. 41; Thacker, *Networks, Swarms, Multitudes*, part two.

13 Brian Massumi, *Parables for the Virtual: Movement, Affect, Sensation* (Durham: Duke University Press, 2002), p. 30.

14 Marianne Pieper and Carolin Wiedemann, "In the Ruins of Representation. Affekt, Agencement und das Okkurente," *Zeitschrift für Geschlechterforschung und Visuelle Kultur*, 55 (2014), p. 66–78.

15 The question then is why use notions such as 'human' and 'non-human' if one wants to overcome exactly this division.

on the swarm, points to the fact that a body is touched by another and mobilized, and this mobilization continues en masse.[16]

Focusing on one example, *Anonymous* (the example that Stalder also chooses)[17], and one specific case, the creation of the LOLcats meme, exemplifies how the concept of affect can help to explain digital swarming. Within the creation of the LOLcats, *Anonymous* corresponds to the basic definition of swarms (directional force without centralized control, more than the sum of its parts).[18] People who were part of these spontaneous, unplanned cat postings, LOLcats, on the online board 4chan explain what happened: they were hanging out on 4chan, had nothing to do, and then somehow got excited by how other users were starting to post cat pictures and still other users were adding sentences to these cat pictures, and without knowing what exactly was going on, they just posted cat pictures as well and started modifying the earlier ones and spontaneously felt part of a community through their common action. More and more users uploaded new cat pictures, and others commented and called these pictures LOLcats. Within a few hours, there were hundreds of LOLcats on 4chan, and these pictures then spread to the whole Internet.

There is already some academic research on LOLcats that qualifies the phenomenon as a meme, even as the most popular Internet meme.[19] The common definition of a

---

16 Eva Horn, "Schwärme – Kollektive ohne Zentrum. Einleitung," ed. Eva Horn and Lucas Marco Gisi, *Schwärme – Kollektive ohne Zentrum* (Bielefeld: transcript, 2009), p. 7–26, here: p. 17.

17 Stalder, *Digital Solidarity*, p. 41.

18 *Anonymous* in general cannot be classified as a swarm; rather, it functions as a "living network," to use a term first employed by Thacker, see also: Carolin Wiedemann, "Between Swarm, Network and Multitude. Anonymous and the Infrastructures of the Common," *Distinktion: Scandinavian Journal of Social Theory* (printed version forthcoming), http://www.tandfonline.com/doi/abs/10.1080/.U1JzvV5vOco#.U4CeRS93aco (last accessed May 24, 2014). But in order to illustrate the concept of an online swarm and to concretize the role of circular affection single actions/moments that are attributed to *Anonymous* are a quite suitable example for the swarm phenomenon – like the LOLcats.

19 The phenomenon of LOLcats and the discussion of memes are interrelated. Most people turn to Lolcats in order to explain internet memes, see: Kate Miltner, "SRSLY Phenomenal. An Investigation into the Appeal of LOLcats," http://dl.dropboxusercontent.com/u/37681185/MILTNER%20DISSERTATION.pdf (last accessed May 24, 2014). Thus memes are defined as digital images, often overlaid with text, animations and sometimes memetic hubs of videos that emerge in a grass-roots manner through networked media and acquire a viral character by spilling over from their 'birth place' into diverse online channels and to other forms of media and thus becoming globally popular. A video that goes viral is not automatically a meme as a meme does not emerge unless people contribute by modifying it, responding to it, and enacting it. This definition by Olga Goriunova refers to the self-reflection of meme culture that platforms as Knowyourmeme have developed by conducting crowd-sourced research into memes, see Olga Goriunova, "The force of digital aesthetics: on memes, hacking, and individuation," English draft which has been translated into German and published in *Zeitschrift für Medienwissenschaft*, 8 "Medienästhetik," 1/2013, draft available online:

meme does not include the term *swarm*, but based on Thacker's notion of the latter, the swarm can be conceived of as the sort of collective that creates the meme. That is, within the emergence of the meme a swarm is at work, and the output of this swarm or these swarms is the meme; in the case of the LOLcats phenomenon, the meme are the cat pictures superimposed with various statements and slogans that then go viral. They only emerged in the interaction of diverse actors on 4chan who did not know each other or realize what the output would be, who did not even intend to create something. Within the interaction on 4chan the single actors become part of something bigger that they can't perceive during its development. That is where processes of affection occur. The *something bigger* is perceivable only after these processes have taken place. The LOLcats became the LOLcats when people started to reflect on what had happened and to speak about the output: they then gave this phenomenon the name *LOLcats* and later identified it as a meme. It was only in the nonintended, unregulated, and ungoverned interaction of the users based on processes of affection that the swarm (directed force without centralized control, more than the sum of its parts) and then the meme emerged.

In contrast to forms of suggestion and imitation from body to body, as described in crowd and collective-behavior theory,[20] in the case of the LOLcats the affective processes do not produce contagious forces but raise the potential of the affected bodies to act and thus are a creative force that has a modifying effect. Furthermore, the affectively inspired interactions have to be situated in relation to a specific online site. Thus in order to understand these processes and to approach the constitution of the digital swarm, the challenge is to understand the specific qualities of the media infrastructure of 4chan, the "meme-factory" as its founder calls it.[21] In contrast to 4chan, other social networking sites – the most frequented of which is Facebook – are also used to spread information, pictures, videos or articles, but no such phenomenon as the LOLcats ever developed there. In order to understand the specific quality of 4chan that inspires the emergence of swarms and memes, I want to analyze it with a new materialist approach.

—

https://www.academia.edu/3065938/The—force—of—digital—aesthetics—on—memes—hacking—and—individuation (last accessed May 24, 2014).

20 For an overview of these theories see: Christian Borch, *The Politics of Crowds: An Alternative History of Sociology* (Cambridge: Cambridge University Press, 2012).

21 Christopher Poole, "Meme Factory," Paraflows 09: Festival, Digital Art and Cultures, Vienna, Austria (2009); see http://digiom.wordpress.com/2010/04/06/moot-on-4chan-and-why-it-works-as-a-meme-factory/ (last accessed May 24, 2014).

3. NEW MATERIALISM AND INFRASTRUCTURE FROM THEORY TO METHODOLOGY

Within the new theoretical movement that affirms the vibrant dynamics and unique capacities of nonhumans, there are strong differences that are expressed in different labels: new materialism, speculative materialism, object-oriented ontology, and actor-network theory. Nevertheless, the common thread linking these approaches justifies our talk about a *turn*: All of the theorists behind the material turn or speculative turn "have certainly rejected the traditional focus on textual critique (...) [and] all of them, in one way or another, have begun speculating once more about the nature of reality independently of thought and of humans more generally," as the introduction to *The Speculative Turn: Continental Materialism and Realism* states.[22] According to this passage, taking into account the agency of nonhuman beings raises the question of access to reality and thus opens an epistemological discussion – and a methodological question. If this is a turn toward "dynamic human and non-human materialities which acquire shapes, operate and differentiate also beyond human perception and discursive representational systems,"[23] then how does one do research from this perspective? How can I approach the infrastructure? Informed by postmodern approaches and theories after the *crisis of representation*, I would concentrate on the discursive level and deconstruct the representations. But can I also approach the infrastructure more directly? The strands of speculative realism and object-oriented ontology have strongly discussed correlationism, "the idea according to which we only ever have access to the correlation between thinking and being, and never to either term considered apart from the other".[24] These strands also assume that anti-representationalist and deconstructivist theories retained correlationism, as the latter would imply a real *out there* that our representations do not meet. In opposition to these postmodern approaches, Meillassoux, one of the most prominent figures associated with speculative realism, tries to find ways to surpass the limits of

—

22 Levi Bryant, Nick Srnicek, and Graham Harman, *The Speculative Turn: Continental Materialism and Realism* (Melbourne: Re:Press, 2011), p. 3.

23 Jussi Parikka, "What is New Materialism-Opening Words from the Event," Blogpost on *Machinology. Machines, noise, and some media archeology* by Jussi Parikka (2010), http://jussiparikka.net/2010/06/23/what-is-new-materialism-opening-words-from-the-event/ (last accessed September 2, 2014).

24 Quentin Meillassoux. *After Finitude: An Essay on the Necessity of Contingency* (London: Continuum. 2008), p. 5.

what we take to be the human standpoint – i.e., finitude – and claims that this takes place through mathematics.[25]

Even though the aim of *neighborhood technologies* is – as outlined in the introduction to this book – to create a transdisciplinary discussion between mathematics and media and cultural studies, I will not try to follow Meillasoux's approach. Neither will I continue to deepen the epistemological discussion on accessing reality. Even these limits align with Meillasoux's approach, though, for he is not concerned about showing that objects are real; rather, he is interested in accessing the material world we are involved in (and thus tends to reject the label *speculative realism* in favor of *speculative materialism*).[26]

In a similar way, Alex Reid, by referring to Bruno Latour, points out: "Once we do away with the modernist's real world, the modernist-correlationist concern of not being able to access it doesn't make much sense."[27] Taking up the viewpoint of Reid and Latour, I do not aim to confront anti-representationalism and new materialism; rather, I want to broaden the poststructuralist background through theories of new materialism that are not in contradiction to anti-representationalism. This includes all of the approaches inspired by Gilles Deleuze and Félix Guattari and their precursors. New materialist methodology can thus focus on relations that surround us, that have material effects that can be traced within human perception, that affect the world which one encounters as a researcher, rather than wondering whether it is possible to approach preexisting matter. Nevertheless, the new materialist methodology that I suggest does not intend "to get away from facts but closer to them, not fighting empiricism but, on the contrary, renewing empiricism" by adding reality to matters of fact, by creating new concepts around the "matters of concern".[28]

Latour's concept of mediators might be conceived of as such an addition: in his theory, "mediators" are all the things that mediate within the socio-technical world, that "transform, translate, distort, and modify the meaning or the elements they are supposed to carry".[29] Because mediators can be analytically conceived as the "things" or "gatherings"[30] (I will come back to this point later) that can transform connectivity into

---

25 See Dolphijn/van der Tuin, *New Materialism*, p. 177.

26 Meillassoux, *After Finitude: An Essay on the Necessity of Contingency*, p. 121.

27 Alex Reid, "Latour and Correlationism," Blogpost on *digital digs. an archeology of the future*, http://alex-reid.net/2013/03/latour-and-correlationism.html (last accessed May 24, 2014).

28 Bruno Latour, "Why Has Critique Run out of Steam," *Critical Inquiry* - Special issue on the Future of Critique, 30.2 (2003), p. 225–248, here: p. 231.

29 Bruno Latour, *Reassembling the Social* (Oxford: Oxford University Press, 2005), p. 39.

30 Latour, *Reassembling the Social*, p. 157.

collectivity, they help to approach the role of the infrastructure in the process of swarm constitution and therefore might help us to understand the subject of this article: the online swarm and its products, the memes. Given that the concept of mediators implies the processes of mediation between persons, artifacts, and symbols, it already points to the model of *occurent relations* within which mediators emerge and work. But much more than Latour, Deleuzian-inspired thinkers such as Massumi stress the concept of relationality.[31] And the latter cannot be conceived without a notion of affect. As described above, the concept of affect can be helpful in explaining the formation of a swarm that creates a meme like the LOLcats. While the concept of the mediators seems important to approach the infrastructural elements at work during the creation of an online swarm, complementing this approach with a notion of affect allows to concentrate more on the relations and thus to focus on what happens within the processes of mediation. "Affects operate in the mode of connectivity, they circulate, create dynamics, produce subjectivity and mobility. They operate as – vivid and dynamic – immanent force".[32]

Such a focus on the dynamic and constitutive relations between bodies or objects, between diverse materialities as suggested by Massumi, implies to take (physical) bodies into account, both human and nonhuman – not as a priori, as something already given, but as being constituted within these processes of circular affection. This assumes a perpetual actualization of the world that stems from connectivity across human and nonhuman, material, and semiotic elements and their "open series of capacities or potencies".[33] New materialist approaches that stress the constitutive role of circular affection and relation can conceptually and analytically reclaim the fundamental reality-making role of materialities without attributing to them a self-contained identity or ontological primacy that would determine other materialities or bodies.[34]

In the following, I will therefore try to approach the elements of the Internet infrastructure as potential mediators in the process of constituting collectivity like the

—

31 Brian Massumi, "Of Microperception and Micropolitics. An Interview with Brian Massumi," *Inflexions: A Journal for Research-Creation.* No. 3. Micropolitics: Exploring Ethico-Aesthetics (2009), http://www.inflexions.org/n3—massumihtml.html.

32 Marianne Pieper, Vassilis Tsianos, and Brigitta Kuster, "'Making Connections'. Skizze einer Net(h)nographischen Grenzregimeanalyse," ed. Theo Röhle and Oliver Leistert, *Generation Facebook* (Bielefeld: transcript, 2011), p. 221–248, here: p. 230.

33 Diana Coole and Samantha Frost, *New Materialisms. Ontology, Agency, and Politics* (Durham: Duke University Press, 2010), p. 10.

34 Milla Tianen, "Revisiting the Voice in Media and as Medium: New Materialist Propositions," *Necsus. European Journal of Media Studies*, Autumn (2013), http://www.necsus-ejms.org/revisiting-the-voice-in-media-and-as-medium-new-materialist-propositions/#—edn18 (last accessed May 24, 2014).

online swarm. First, I will briefly describe the interface of 4chan and the options for its use, the frameworks for action within which the "abstractions of algorithmic measures" return to bodies, and then I will try to grasp their "affective dimension".[35] In order to understand the abstractions of 4chan, what it does and how its use is structured, I have visited and also tried to use it regularly since the beginning of 2009.[36]

4. 4CHAN AND ITS FRAMEWORK FOR ACTION

Compared to studies on Facebook and other Web 2.0 sites, there are relatively few works that study 4chan.[37] That seems surprising, as 4chan attracts around 18 million individual monthly visitors and generates 1 million posts per day;[38] it is also the birthplace of most of the popular Internet memes and of Anonymous, the phenomenon that grew from 4chan to a globally active hacktivist collective. But visiting 4chan is disturbing: the content – especially on the /b/ board, the most active section – is often very sexist, racist, and homophobic, and even child pornographical.[39] I will not elaborate on these aspects further; rather, I will first concentrate on the *framework for action*.

—

35 Jussi Parikka, "Across Scales, Contagious Movements," Blogpost on *Machinology. Machines, noise, and some media archeology* by Jussi Parikka (2010), http://jussiparikka.net/2010/05/04/across-scales-contagious-movement/ (last accessed May 24, 2014).

36 I responded to some posts and participated in the conversations.

37 Some of the research on 4chan concentrates on the topic of anonymity and ephemerality which is possibly facilitating the creation of memes and important references for my analyses., i.e. Lee Knuttila, "User Unknown: 4chan, anonymity and contingency," *firstmonday. Peer-reviewed journal on the internet* 16/10 (2011), firstmonday.org/article/view/3665/305 (last accessed May 24, 2014); Jana Herwig, "The Archive as the Repertoire. Mediated and Embodied Practice on Imageboard 4chan.org," ed. Günther Friesinger, Johannes Grenzfurthner, and Thomas Ballhausen, *Mind and Matter. Comparative Approaches Toward Complexity* (Bielefeld: transcript, 2011), p. 39–56; Gabrielle Coleman, "Our Weirdness Is Free," ed. David Auerbach and Gabrielle Coleman, *Here Comes Nobody. Essays on Anonymous, 4chan, and the other Internet Culture*, 15 (2012), http://canopycanopycanopy.com/15/our—weirdness—is—free (last accessed May 24, 2014).

38 Luke Simcoe, "The Internet is Serious Business. 4chans /b/ board and the Lulz as Alternative Political Discourse on the Internet," (MA Major Research Paper, Ryerson University – York University, 2012),http://lukesimcoe.tumblr.com/post/47982246987/the-internet-is-serious-business-4chans-b-board-and (last accessed May 24, 2014), p. 2.

39 In a sample post from 4chan's /b/ board that Luke Simcoe cites this is described more precisely: "This isn't a family-friendly site. You'll see lots of nigger dicks, girls with electronic gadgets in their asses, who look 13, racism, sexism, retards, faggots, heads split open, dead cats, and TONS of shit you won't understand ... oh and here's a video of a guy being murdered with a hammer," see Simcoe, "The Internet is Serious Business," p. 8.

The main difference between 4chan and other Web 2.0 sites is that it does not require usernames or passwords, does not archive content or track users. The site and its information architecture is copied from available forum software; it is composed of boards, threads, and posts. Each board is themed (e.g., /v/ is *Video Games*), and the posts within the boards are classified into threads. Posts starting a thread have to include an image, while images in replies are optional. As there is no registration and no login necessary to start posting, from a technical point of view accessing the site and participating seems as easy as it can be.[40] While other Web 2.0 sites pull out all possible information from profile entry fields and postings in order to create data-based representations of the individual users (the individual's social graph) and of the constant communications (the individual's timeline, activity stream) in order to understand the relations between users (social graphs), on 4chan users do not leave any of this.[41]

In 4chan's communication options this element of anonymity is even taken a step further. Unlike on other sites, where being anonymous usually means not to register under a real name or identity but creating a pseudonym and sometimes also an account with this pseudonym, on 4chan there are no accounts. All information is entered on a per-post basis. Furthermore, creating a nickname does not guarantee individuality, as the same nickname is not blocked for other users. If a user claims a pseudonym, any other user can claim it for herself in any following posting. And if a user, as a consequence, refuses to create a nickname, 4chan automatically gives him or her the name *anonymous*. As an effect, if all users are represented as *anonymous*, nobody will know who has been talking. Furthermore, because of the structure and design of the postings, one cannot tell how many users are answering one another in a thread. Instead of user IDs, a unique identifier is attached sequentially to every post on the board, which allows users to reply to a given post and creates a chronology of posts. Thus, as Jana Herwig points out, users respond not to other users but to opinions and imagery.[42]

However, the users are not just affected by the ideas. The interplay of ideas with and within this kind of infrastructure differs from other Web 2.0 sites not only in allowing anonymity but also in offering ephemerality and a specific sort of contingency. Unlike any other social networking site, 4chan has no memory. All postings that do not generate responses will automatically move down the queue and will eventually

—

40 But it is not only the technical codes that could exclude some people but also cultural codes that have been established on 4chan, see Wiedemann, "Greetings from the Dark Side of the Internet".

41 Herwig, *The Archive as the Repertoire*, p. 52.

42 Ibid.

be deleted. Therefore, only the posts that gather enough users are repeated and stay on top of the sites that the boards are organized in and thus can attract more and more attention within a few seconds. In the article "4chan and /b/: An Analysis of Anonymity and Ephemerality in a Large Online Community," MIT researchers collected data for two weeks, compiling 576,096 posts in 482,559 threads.[43] They found that the median thread spends just five seconds on /b/'s first page before being pushed off by newer posts and that the median thread expires completely within five minutes.[44]

The dynamics on 4chan are further strengthened by the fact that statistically very few users view the same page at the same time, since the site does not update in real time (users must reload to see something new). The combination of these structural elements and 4chan's ephemerality and anonymity engenders contingency. Knuttila points out that 4chan memes are a "reaction to, and embodiment of, contingency".[45] The same is true for the swarm emerging on 4chan: on the one hand, the swarm exists as a collective within the creation of the meme, thus within the circulation based on circular affection and thus on constant modification. Therefore it stands simultaneously *against* and *for* ephemerality, anonymity, and contingency: the emergence of a swarm on 4chan is "neither necessary, nor impossible," because "the anonymous interface creates infinite possible interactions".[46]

5. AFFECTS, CONTINGENCY, AND RELATIONAL EVENTS

Having approached the infrastructure as a framework, I will now try to grasp the affective dimension and thus concentrate on what happens between the frames and the bodies on 4chan – between the things, the emergences, that also partly constitute the frameworks. The emphasis on contingency on 4chan (the notion that swarms could emerge as if they were completely accidental) recalls Massumi's emphasis on the term *relational event*. Massumi considers *relation* the first key term for defining the notion of affect. As for *event*– as in, "It's all about event" – it describes the occurrent relation that encompasses affect, and thus in his words the "relational event will play out differently every

—

43 Michael S. Bernstein et al., "4chan and /b/: An Analysis of Anonymity and Ephemerality in a Large Online Community,"", Association for the Advancement of Artificial Intelligence, projects.csail.mit.edu/chanthropology/4chan.pdf (last accessed May 24, 2014).

44 Bernstein, "4chan and /b/," p. 56.

45 Knuttila, "User Unknown".

46 Ibid.

time. [...] The region of occurrent relation is a point of potentiation. It is where things begin anew".[47]

This understanding of affect could indicate that the affective processes are completely contingent. But this interpretation would be too simplistic and would not fulfill the concept's theoretical value. When Massumi introduces "tendency"[48] as the second key term for understanding affect, it becomes obvious that this concept is neither asocial nor ahistoric. Applying it means taking into account "presuppositions"[49] and considering the tendencies of the bodies activated via the affect. On 4chan as in any other sphere that is created and used/inhabited by social beings, there are presuppositions. The bodies interacting on 4chan have a past; the bodies have tendencies as to what extent they are activated through processes of affection. Herwig, who has also studied the content of the postings on 4chan, has proved that there are specific codes establishing unspoken rules as to which posts attract attention.[50] But the concept of affect allows us to consider a precise "activation not only of the body, but of the body's tendencies".[51] And that is what happens on 4chan: Consistently, memes have emerged that were not predictable in their specific actualization.

Presuppositions are not determining/determinative, and that is exactly because there are affects at work. Applying a concept of affect opens up the perspective on "the world stirring,"[52] the continuous processes of subject constitution within a crowded and heterogeneous field of "budding relation".[53] The strong ephemerality and the specific contingency on 4chan are important for understanding the affective dimension of the board, as they affect the dynamics as well as the content of the postings, which again is important for the processes of circular affection. The knowledge that those posts/threads which do not attract attention will disappear so quickly might inspire users to (a) start threads with explosive, irritating posts, or pictures or statements that interrupt the stream, (b) modify their threads if they get deleted too quickly and start experimenting with diverse threads, excited by observing how other users react to them, wondering which one survives more than a few seconds, and (c) start as many threads as possible as quickly as possible in order to increase the chance that one post stays on the site

---

47 Massumi, *Of Microperception and Micropolitics*, p. 2.
48 Ibid., p. 3.
49 Ibid., p. 3.
50 Herwig, *The Archive as the Repertoire*, p. 50.
51 Massumi, *Of Microperception and Micropolitics*, p. 3.
52 Ibid., p. 4.
53 Ibid.

at least a little while. All of these three dynamics facilitate processes of circular affection. Furthermore, the fact that there is no automated update in real time by the site itself means that the user replying to a post within a thread, or typing or uploading a picture, will not yet be able to see whether many other users have already replied to the same post and whether his or her reaction could be outdated. This structural condition also accelerates the interactions and thus the frequency of *micro-shocks*.

Affections are inspired by *micro-shocks*, as Massumi, claims – by little interruptions that are in every shift of attention, "for example a change in focus, or a rustle at the periphery of vision that draws the gaze toward it".[54] These shocks can pass unnoticed; they are felt but are not registered consciously. They are registered only in their effects. If 4chan users continuously click the update button and always see new content appearing on the site, there are permanent micro-shocks at work. Their bodies are in motion, scrolling and saving images to their hard drives, intertwining them back into circulation, creating novel symbols and forms.

And within these micro-shocks, within these affective processes, relations between the users develop. According to Massumi subjects emerge within the occurrent relations that are based on reciprocal affects.[55] The 4chan user becomes a subject as part of the collective, and the collective itself emerges as a subject, as a swarm – with the meme being another emergent subject in these processes of subject constitution inspired by a board like 4chan. Moreover, this board, which functions as a network as it connects diverse actors, does not have any existence beyond these permanently occurring relations. If there are no users interacting, then there is no network. 4chan as a network only emerges as a system of dynamic relations, in the affective processes.[56]

The specific anonymity of 4chan lowers the restraints that could control the reactions to these affects. Users are not assumed to represent *themselves*, as they are on other social media sites, but can interact without being watched. They can experiment without being surveilled.[57] The experimental frame is strengthened by the almost inevitable ephemerality of all interactions, of the abundance of existing relations. Herwig regards 4chan as a laboratory "where body-subjects collectively explore meaning," as a dance

---

54 Ibid., p. 3.

55 Ibid., p. 3.

56 Wiedemann, "Between Swarm, Network and the Multitude".

57 For an overview of the discussion on the relation of anonymous infrastructures and the notions of accountability and trust, see for example Batya Friedman, and John C. Thomas, "Trust Me, I'm Accountable. Trust and Accountability Online," Summary of a panel discussion with experts (1999), http://vsdesign.org/outreach/pdf/friedman99accountabilityonline.pdf (last accessed May 24, 2014).

that "transgress(es) the borders between mind and matter, media and body, information and perception, which the dominant discourse has so meticulously established".[58]

4chan recalls Erin Manning's and Brian Massumi's experiments in the *Sense Lab* in Montreal. The SenseLab is a "laboratory for thought in motion" where "punctual research-creation events supplemented by ongoing, year-round activities" are organized.[59] The Sense Lab creates space for encounters that are not meant to produce sense but are instead playful – encounters of bodies that bump into each other without any meaning.[60] On 4chan the movements on the screen are not directly perceivable as gestures by bodies; rather, they are movements of letters, images, text – of instruments that one would actually assign to the sphere of representation. But on a site like 4chan they leave the representational frame – they are not archived, they become ephemeral – and thus text on 4chan becomes movement, meaning becomes fluid, and images evaporate like gestures.

With the Senselab, Massumi and Manning had the ambition to design constraints "that are meant to create specific conditions for creative interaction where something is set to happen, but there is no preconceived notion of exactly what the outcome will be or should be." There is "no deliverable. All process."[61] Similarly, on 4chan there is no prior agenda as to what should be posted. In addition, such an online board, unlike an urban space, has no borders and no center. Hence, theoretically there is an even higher chance for heterogeneous encounters, as people who do not know each other and neither see nor hear each other can assemble and communicate, and all the communicative acts theoretically can be equally perceived. On this online playground the fluidity of representations and the impossibility of individual accounts facilitate the formation of digital swarms, of swarm creativity.

6. 4CHAN AS AGENCEMENT (AND THERE ARE MANY QUESTIONS LEFT)

A further new materialist investigation of such infrastructures could take up three remaining endeavors. Indeed, all three endeavors would comprise a further mapping of the field in which these infrastructure emerge, but nonetheless, I want to delineate some

—

58 Herwig, *The Archive as the Repertoire*, p. 52.

59 See: http://senselab.ca/wp2/ (last accessed May 24, 2014).

60 Massumi, *From Microperception to Micropolitics*, p. 13–15.

61 Ibid., p. 15.

thoughts on each of these possible aspects of inquiry: First, it should analyze the infrastructure as a *gathering*, as Bruno Latour suggests,[62] which for example means to further investigate how this infrastructure is produced. Understanding 4chan as a gathering involves pointing to additional materialist questions. That means also asking *old* materialist questions, that is, in the tradition of historical materialism. Even if not following a Marxist notion of conscience, within biopolitical capitalism questions of ownership and exploitation are still relevant. In the entanglement of hard- and software there are various materialities at work, from practices of labor to production chains and onto the chemicals and components that comprise the technology.[63] Beyond these broader geopolitical and ecological issues, there are still further materialist questions at hand, for instance if one focuses on 4chan's software level: The site belongs to 4chan's founder, Christopher Poole. It is not a non-profit site but is dependent on advertisement. There are moderators who are instructed by Poole, and these moderators can delete content. The infrastructure is static, thus the users cannot modify it. There is no open code. In the face of such facts, further essential questions would include the following: Is there any relation between the advertisement and the memes? Which content gets deleted? And how do you become a moderator?

Secondly, this investigation would have to situate the infrastructures in a broader context, in order to understand their integration in power structures and hegemonic discourses. Given that 4chan – and, likewise, the swarms and memes that emerge on it – would not exist if the board was not used and that this kind of living network is consistently constituted by people posting on it, the broad acceptance and popularity of such a site has also to be contextualized. With which specific social, political, and economic assemblages are both these technologies and techniques of swarming interrelated?

And third, the inquiry would have to figure out how to reconfigure a notion of emancipation or rather subversion within the current situation with regard to the previous analyses. According to Sebastian Vehlken's account of swarming as a cultural technique, swarms have become relevant as structures of organization and coordination, as effective optimization strategies and zoo-technological solutions for "the governmental constitution of the present itself, in which operationalized and optimized multitudes have emerged from the uncontrollable data drift of dynamic collectives".[64] The logic of

---

62 Latour, "Why has Critique Run out of Steam," p. 233.

63 Jussi Parikka: "Dust and Exhaustion. The Labor of Media Materialism," *Ctheory*. October 2, 2013, http://www.ctheory.net/articles.aspx?id=726 (last accessed May 23, 2014).

64 Vehlken, "Zootechnologies," p. 127.

contagion, which is linked to the mathematics of epidemics and organization theories, becomes a key tactic in commercial, security, and technological contexts within current forms of capitalism. As Tony Sampson notes, the notion that spontaneous collective moods can be guided toward specific goals seems to be the latent exercising of an affective biopower over an increasingly connected population.[65] This governmental constitution, which is exhibited, for example, in viral marketing, tries to capitalize on affectability, or the users' capacity to affect (and to be affected). An infrastructure like 4chan which creates a dynamic informational space "suitable for the spread of contagion and transversal propagation of movement (from computer viruses to ideas and affects),"[66] can serve biopolitical and capitalist vectors, but – as I have tried to show – also opens up a space for encounters that are neither completely controllable and nor exploitable.

If current forms of biopolitical capitalism govern via computed evaluation and prediction in order to capitalize each aspect of life, then unpredictable and non-exploitable emergences as well as anonymous cooperation beyond individual pursuits of profit indeed hold a subversive potential. An infrastructure like 4chan enables practices of sharing or collaboration that surpass any system of recognition – nobody knows who contributed what. Hence, the infrastructure enables spontaneous common creations. Affect circulation and swarm formation on 4chan may neither automatically inspire nor involve the formation of a political agenda with common goals which would indicate "a culture of digital solidarity,"[67] but such an infrastructure offers the potential for experimentation and experiences of creative interaction, for becoming collective in new ways.

7. EMERGING NEIGHBORHOODS

The swarms that emerge on 4chan can be identified as techno-social groupings and thus as neighborhoods. But in contrast to classical concepts of neighborhoods in the case of Anonymous, anonymity and spontaneity are constitutive for the swarms. These neighborhoods do not exist beyond the processes of circular affection through the anonymous postings and their specific dynamics. Instead they only emerge as systems of dynamic relations, in the affective processes. They do not share a common background, they are

—

65 Tony Sampson, *Virality: Contagion Theory in the Age of Networks* (Minnesota: University of Minnesota Press, 2012) p. 126.
66 Tiziana Terranova, *Network Culture. Politics for the Information Age* (London: Pluto Press, 2004), p. 67.
67 Stalder, *Digital Solidarity*, p. 14.

not based on common myths and narratives, they have no past and thus no acquaintance. They do not need any beforehand-shared identity that is based on a common story of a neighborhood in which people live and organize their lives next to each other. Only within the moment of the encounter, of the circular affection the elements are constituted as neighbors.

Nevertheless these new neighborhoods are *local*: Analyzing occurent relations within and on 4chan does not only point to the fact that people who do not meet physically but are connected via a technical medium can unexpectedly perceive a sort of community. It rather highlights the fact that these emerging neighborhoods include media technical elements which themselves become essential parts of the relational events without previously being attributed such a meaning. People becoming part of the swarm might be dispersal but their neighborhood is localized as it is based on a specific infrastructure. A technologically informed notion of neighborhood concepts takes into account the constitutive role of media technologies – as exemplified in the analyses of 4chan from a new materialist perspective. Such a notion further inspires a theoretical discussion of the relationality of the concepts of locality and materiality that situates the technologies as well as their analyses within specific dispositives.

GABRIELE BRANDSTETTER

CHOREOGRAPHING THE SWARM

RELATIONAL BODIES IN CONTEMPORARY PERFORMANCE

ABSTRACT

In recent times, the swarm model has become a prominent concept for the portrayal of collective movements and of transitional assemblies both in social and in media-controlled spaces ranging from surveillance cameras to google maps. Media theories exploit the swarm metaphor for concepts of locally organized political and social neighborhoods. Swarms are thus brought into play as a template in order to better comprehend phenomena like Internet-based *smart mobs* and other non-hierarchical forms of participation.

This article asks which structural principles of movement are characteristic for swarming. How can swarms or flocks be described as choreographic processes? By considering examples from contemporary dance and performance, the article examines which decisive principles of proximity/distance and of cohesion in movement are employed, and contemplates the applied kinesthetic processes and impulses of control.

Where are the boundaries of *choreographing* the swarm? How are they situated? And how can the different modes of (passively) participating in the movement of a swarm and of (actively) observing the actual – but never completely determined – figurations of swarming come together in performance analysis?

1.

This paper considers the question how choreographies and performances of *swarms* and of *swarming* show neighborhood-effects. Are the relations between swarm-participants induced more by media and/or by body-techniques? Within the conceptual framework of performing arts and choreography, the dynamics of collectives and the movements of swarms can be categorized in different patterns. In the following, I concentrate on two modes of neighborhood-effects in swarm-movement: The mode of *coherence* on the one hand, and the mode of *dispersal* (Zerstreuung) of bodies on the other. I will give evidence to both of theses modes of relating bodies – the dynamics of cohesion and the techniques of remaining related *within* the movement of dispersal – by some seminal examples of swarm-choreography.

Before that, I give a short outline of theories of participation and body-synchronisation in the field of performance, and discuss how they take advantage of or question swarm concepts.

My first example is the well-known phenomenon of a *flash mob*, a public happening which oscillates between political, media and art performance. One of the first flash mobs dates from July 2003, when about 250 people gathered at New York's Grand Central Station before proceeding to the nearby Grand Hyatt Hotel. There, they assembled in the gallery in a calm and decorous manner. At exactly 7.12 pm, they burst into thunderous applause, which lasted for 15 seconds. After that, the crowd quickly dispersed, while police cars drew up outside with wailing sirens.

The intriguing aspect about flash mobs or smart mobs in comparison to more traditional forms of gatherings lies in their utilization of media-based applications or websites for synchronization. Referring to Howard Rheingold's theses on smart mobs[1], one could ask whether and in what ways the sudden emergence of temporary, dynamic communities can be conceived of in terms of a swarm model. What conclusions can be drawn from such forms of synchronizing human gatherings for modes of participation, if one sees the organizational form of the swarm – the non-hierarchical, self-regulating order, the temporal, rhythmical and collective spatial formation, the coherence and dissolution of the association – as a model of collective action?

The above example still refers to traditional notions of participation since the flash mob constitutes a network of participants in real-life activity. This collective activity produces an assembly which seems to behave like an *audience*. However, the applause is not directed to a performance or to its actors, but *itself* is the performance. The visible and audible live performance exhibits the aforementioned synchronization effects of a fortuitous, temporary collective. "Participation" therefore literally means nothing more than to be a part – in contrast to be a part of a gathering which comprises a purpose. One of the anonymous organizers of the flash mobs and their applause commented: "It is wonderful to take part in something so unexpected."[2] Recently, in media theories of the Internet, the word *sharism* has been attributed to phenomena like flash mobs and

---

1 See also Howard Rheingold, *Smart Mobs – The Next Social Revolution* (Cambridge/Ma.: Basic books, 2002).

2 See also introduction by Gabriele Brandstetter, Bettina Brandl-Risi, Kai van Eikels, Ulrike Zellmann "Übertragungen. Eine Einführung," ed. Gabriele Brandstetter, Bettina Brandl-Risi, Kai van Eikels, *SchwarmEmotion: Bewegung zwischen Affekt und Masse* (Freiburg i. Br.: Rombach, 2007): p. 7–61.

digital communication locations.[3] And in the course of the last decade, media formats like blogs, wikis, and platforms like Facebook, Twitter, or MySpace constantly stirred up a discourse about new concepts of participation, and new forms of communities.[4]

The main focus of my inquiry lies on the transfer of movements between individuals and collective forms of actions: How can we describe the genesis, the transfer and the dynamics of such neighborhoods – in terms of mediality and corporeality, and in categories of relations: of placement and displacement, proximity and belonging, and movement and empathy.[5] How can we understand processes and rhythms of action, movement and perception in terms of synchronization so as to analyse their significance for participation in collective (social and artistic) processes? And how is the synchronization of individuals and collectives organized into transformative procedures?[6]

2.

As mentioned above, swarms have become not only a popular metaphor, but also a model that exemplifies different social and artistic performances of movement synchronization. In recent years, the model of the swarm has achieved the status of a paradigm in the social and cultural analysis of collective movements. I am interested in this shift of paradigm within models and theories of participation in theatre and performance, and want to know how to read the structures of swarm movements and their performativity.

What are the principles of swarm dynamics and their movements, how do they achieve their status as a model of collective synchronisation? These structures of flow and movement dynamics are instructive for the following questions of the participatory aspects and of the possibility of choreographing the swarm. Thus, in the next sections, I will briefly discuss some key aspects of swarm phenomena.

Swarm dynamics are based on three simple local rules of movement:

—

3 Issac Mao, *Sharism: A Mind Revolution* http://freesouls.cc/essays/07-isaac-mao-sharism.html [retrieved 03.09.2014].

4 On questions of swarm-like principles within networks and the related discourses see also Sebastian Vehlken, *Zootechnologien. Eine Mediengeschichte der Schwarmforschung* (Zürich: diaphanes, 2012): p. 409–414.

5 See also Brandstetter, Brandl-Risi, van Eikels, eds., *SchwarmEmotion*.

6 See also Kai van Eikels, "Diesseits der Versammlung. Kollektives Handeln in Bewegung: Ligna, Radioballett," ed. Brandstetter, Brandl-Risi, van Eikels, *SchwarmEmotion: Bewegung zwischen Affekt und Masse* (Freiburg i. Br.: Rombach, 2007): p. 101–124.

"Move in the direction of the centre of those nearby."

"Move away as soon as someone comes too close."

"Move in the average direction of your neighbours."[7]

These local movement rules already lead to the complex collective movement phenomenon that has recently become more and more topical in various areas like biology, economy, media and the arts. The projection of the swarm phenomenon onto various cultural and social phenomena – from football teams and global migration movements to the formation of subversive groups – shows that literal transfers of the term and phenomenon of swarm are taking place.[8] In a good part of these projections, *swarm* functions more as an umbrella term than a critical technical term. However, current tendencies in the adaptation of the swarm concept in the media, in marketing and communication networks show that the minimal definition of swarm movement formulates basic rules of adaptation and direction which constantly re-define the relationship between individuals and the crowd as a unit. Simultaneously, these instructions for the control of swarms reveal that the swarm phenomenon cannot be defined or depicted with exactitude. The very fuzziness und unpredictability of swarm movements is a crucial structural element of the performative organisation of swarms.

The following thoughts revolve mainly around questions of the dynamics of swarms, the organising function of rhythmical movement, and the formation, synchronisation and dynamic figures of collectives – structures of or as a *choreography*. The main focus lies on the complex relationship between inside and outside perspectives on swarms.

3.

Swarms appear as rhythmically moving formations whose elements themselves are in continually changing movement relationships with one another: physical relationships that are mobile and embedded in a collective overall movement. Swarms are constantly moving, with constantly changing speed, dynamics, rhythmical figuration, shifts of

---

7 Excerpts from the concept for the 10th German Trend Day (Hamburg, 2 June 2005): *Schwarm-Intelligenz. Die Macht der smarten Mehrheit* (engl.: *Swarm Intelligence. The power of the smart majority*) http://www.trendbuero.de/trendtag/index.php?f—CategoryId=7&n=1 [retrieved 08.05.2005].

8 Michael Gamper, "Massen als Schwärme. Zum Vergleich von Tier und Menschenmenge", ed. Eva Horn, Lucas Marco Gisi, *Schwärme – Kollektive ohne Zentrum. Eine Wissensgeschichte zwischen Leben und Information* (Bielefeld: Transcript, 2009): p. 69–84.

direction, distance from members one to another and spatial organisation (within a finite range that is not too large). If the change of speed or direction is too abrupt and wide-ranging, the swarm formation breaks up. The *coherence* of swarms is therefore one of the most fascinating aspects: the *rule* to follow the direction of the majority, to keep moving constantly, to head for the centre while keeping close to and distant from other swarm members functions as a general structural principle. According to each type of swarm, the coherence and the *identity* of a swarm collective is brought about in differing ways. Moreover, swarm researchers have found that reaction times *within* the swarm are many times shorter than the individual reaction time of a single animal.[9] The findings of swarm research in biology[10], synergetics[11] and complexity theory have lead to a discussion of the swarm model as an organisational model also in anthropological, sociological, political and economic science. Behavioral scientists at Leeds University[12] developed a computer program that attempts to describe the behavior of bird flocks without being based on determinist premises. It shows that there are three zones surrounding each bird in a flock: an outer zone (attraction), an inner zone (repulsion) and an orientation zone (cohesion). The latter is produced by an overlap of the first two zones, those space-body segments constantly re-balance in the process of movement and in the variation of dynamics. Examinations like these of a time-space model of swarm movement attempt to initially factor out determinist explanations of swarm behavior, e.g. premises of evolutional biology. What is remarkable here is the tendency to a holistic view; the tendency to regard swarms as units, as one giant organism.[13]

The observation of swarms (in this case, bird flocks)[14] makes visible many salient features of the phenomenon. Swarms can be described as groups of individuals or single members of the same species, who are self-organised by means of movement communi-

—

9 See also "Reisegruppen. Gesellige Zugvögel," *geoscience-online.de, Das Magazin für Geo- und Naturwissenschaften* http://www.g-o.de [retrieved 24.11.2004].

10 See also Kevin Kelly, *Out of Control: The New Biology of Machines, Social Systems, and the Economic World* (New York: Basic books, 1995).

11 See also Arkady Pikovsky/Michael Rosenblum/Jürgen Kurths, *Synchronization: A Universal Concept in Nonlinear Sciences* (Cambridge/UK: Cambridge University Press, 2003).

12 Jens Krause/Iain Couzin, "Collective Behaviour in Animals," Tilman Küntzel, ed., *Stare über Berlin. Ästhetische Analogien des Vogelsangs* (Saarbrücken, Pfau Verlag, 2004): p. 24–29.

13 On questions that started to be discussed in culture studies, social studies and media theory see also Sebastian Vehlken, *Zootechnologien. Eine Mediengeschichte der Schwarmforschung*.

14 Compare Rudolf zur Lippe, "Die Mauersegler der Descalzos," *Scheidewege. Jahresschrift für skeptisches Denken* 35 (2005/2006): p. 209–225; see also id.: "Bei den Kranichen im Linumer Bruch," *Scheidewege 36* (2006/2007): p. 378–396.

cation and who act jointly without central control. One of the – biological, or economic – advantages of this organisation is the increase in efficiency, for instance in reaction speeds, the ability to keep forming continually and to act flexibly and in co-ordinated fashion without prior planning. There are these patterns of flexibility, synchronisation of actions and self-organisation, which make the swarm a model for organisation-theory. In his volume *Out of Control*, author Kevin Kelly contemplates the potential of the rhythmical patterns of swarm-movement synchronisation, which is more flexible, creative and autopoetic than directive and hierarchical systems. And Mark Granovetter coined the term of *weak ties* in small world networks and related this to the synchronisation of members of certain social groups and networks – of collectives which resemble the contingent coherence and non-permanent structure of a swarm. Howard Rheingold advanced the buzz word of smart mobs, transferring the swarm model to contemporary mobile technologies of communication (via internet, cell phones, and other wireless synchronisation of movements).[15] In his adaptation of the swarm phenomenon to social, economic and organisational processes he claims "highest profits through swarm intelligence"[16] and promises significant transformations in business models, economy and even society: "You can only profit from swarm intelligence if you are part of the whole."[17] And this participation – coordinated through communication media – leads to one of his key postulates: "The majority is smarter than each of its members."[18] The downright (socially) utopian idea pertinent to the model of smart mobs – that is, of a community which is not organised in institutions, associations or trade unions – confers an emphatic notion of swarms to social, economic and media processes: as prospect of a »social revolution«.[19]

The apparent mystery of swarm intelligence is their emergent, flexible, non-hierarchical self-organisation. It is particularly this aspect that is transferred as a model to economic, military and media technological control programs.

---

15 See also Rheingold, *Smart Mobs*.

16 10th German Trend Day, *Schwarm-Intelligenz. Die Macht der smarten Mehrheit* (engl.: *Swarm Intelligence. The Power of the Smart Majority*) http://www.trendbuero.de/trendtag/index.php?f—CategoryId=7&n=1 [retrieved 08.05.2005].

17 Ibid.

18 Ibid.

19 Ibid.

4.

In terms of a theory of performance and movement it is interesting that – with the swarm model – this moving entity is neither shaped as a collective by the *intention* of individuals, that is, by questions of regulation and controllability – nor is it simply explicable by cybernetic processes. Rather, swarm dynamics evoke approaches from complexity theory and autopoiesis. Micro- and macro-structures of communication overlap in their formation and de-figuration in movements and as movements. For the observer, the *fascination type* of the swarm is characterised by a paradox form(ation) of movement. The individual movements – in the throng of the bodies and the whirlwind of the overall dynamic – seem to be erratic or chaotic. For the observer, there is the pressure of being swept along by the movement (actively or emotionally). And at the same time, from the captivated observer's outside perspective, the swarm appears as an entity that seems to be regulated as if guided by an invisible authority.

However, it is important to note that both an inside and outside view of the swarm cannot be taken *simultaneously*. The rule-driven irregularities captivate the gaze and let micro-structures and partial figures emerge by the processing of multiple parallel local interactions. This results in the unpredictability of a change of direction or of a collective loop. These patterns are set up elusively in time and space. This paradox configures time both as a simultaneous and a consecutive 'unity' as swarms create simultaneous shapes in continuous re-formations. It is a formation as a collective which does not derive its control from the act of *orientation*, and which is not formed by causal or functional, directional movements. The interaction of individual bodies within the swarm occurs in a rhythmical figure in space, which is kinaesthetically produced as a *felt environment* and appears as a swaying, roaming pattern.

5.

The search for the *meaning* of swarm phenomena and the search for organisation criteria determine parts of the scientific as well as the aesthetic perception of swarms. Metaphors for the order of swarm formations, figurative images of *military formations* and of swarms – again and again such comparisons lead to the interpretative field of anthropomorphism, projecting human forms back onto non-human forms. *Hermeneutics* of bird flocks and other swarm formations date back to antiquity, to the analysis of bird migration (cranes) as prophecy and signs of fate. Evidence as history of the *prevision* of fate in bird flight (which, conversely, is characterised by its very unpredictability) goes by structures and units of meaning, and so does the aesthetic perception of the swarm. The beauty of the swarm's movement, its hypnotic, transcending potential for fascination lies in the desire to discover ordered figures, ephemeral structures and images of bodies in space within the eddies of the incalculable. Here, the dynamics of the collective play with a limit that marks an overspill, (literally) overshooting this boundary in an instance of extravagance and luxury.[20]

The phenomenon of an irregularity that is constantly ordered through movement – the swarm as dynamic shape, which is at best comparable to the ephemerality of clouds (similarly inviting the reading of meaning or images) while simultaneously being intangible in its multipartite nature – seems to elicit ways of decoding and interpreting as well as giving rise to the creation of computer simulating models in art and the media. The lack of directional movement, or, in other words, the non-intentional, is a key characteristic of swarm dynamics and their aesthetic appearance. It is not an assignment or a causal chain of processes which determine the movement. The complexity of the shape's movement does not follow a pattern of stimulus and reaction, and overall the open, mobile con-figuration of the swarm cannot be described as action or re-action, but as a form of *movement as answering*. The performativity of swarm movement as overall figuration is therefore not controlled by individual acts. Rather, it is a phenomenon of *emergence*. Swarm dynamics do not just have qualities of emergence, such as the unforeseeable, the unpredictable or the non-controllability of the process by the subject and its instances of control. Swarms virtually appear as *embodiments* of the phenomenon of emergence.

6.

How can we understand the movements of swarming – beyond the organisation and directing of individual and collective movements by prescription of body actions, by notation or repetition? How can we describe the dynamics of coherence of bodily neighbourhoods in dance?[21]

---

20 According to Georges Bataille a swarm suggests a potential of transgression and excess, i.e. dynamics, in contrast with the economy of the division of labor. See also George Bataille: "Eroticism," Fred Botting and Scott Wilson, eds., *The Bataille Reader* (Oxford: Blackwell 1998): p. 221–274.

21 See also Gabriele Brandstetter, "Swarms and Enthusiasts. Transfers in/as Choreography", *parallax 46: installing the body* (2008): p. 92–104.

Swarming in group and mass movements denotes the dissolution of fixed formations into open groups whose coherence and structure are confusing and seemingly uncontrollable. In the field of the military semantics of *swarming* – raised by Virgil as well as Luther and Grimm – the swarm movement corresponds to the model of partisans, in contrast to the military order in the uniform formation of the *corps*.[22] In this, a very particular feature of the swarm comes to the fore that makes swarm phenomena exciting, especially with regard to concepts of movement in art and aesthetics: swarms embody *movement beyond mimesis*. The movement of a swarm cannot be imitated or repeated in exactly the same way. It may well be possible to simulate swarm phenomena, for example with a computer; and swarm effects can be induced and – to a certain degree – be set up. But swarms elude the mimetic reproduction of (physical) performance and of their emergent appearance in time and space. There is an (apparent) paradox or contradiction in the sense of imitation and reproduction. The form and formation of movement, which can be staged by means of (physical) discipline and choreographic orientation, usually has structural characteristics that are contrary to the emergent non-control in a swarm movement. For instance, the *mise-en-scène* of a group and mass physical movements – as a broad definition of choreography – is mostly governed by rules concerning rhythm, time and space, and marked by a time frame with a clear beginning and end. Further, in terms of its process and structure it is vulnerable to disturbances or transformable by chance disruptions. In swarm phenomena, however, the beginning and the end are difficult to identify, as are disruptions of the continuous and continuously self-transforming movement (even when disturbed). The movement of the participating individual bodies blends into the swarm configuration not by dint of imitation, but rather through para-mimetic transfer processes.

In this context, the following questions arise: Is it even possible to create swarms artificially – to produce them as *art*? In what ways can these emergent, complex movement-collectives function as models for choreographic work? Can swarms represent a performative model for the relation between individual and group, for the coherence of moving groups and the conditions of their co-ordination in time and space? Finally, can they be a model for an emergent unfolding of form that proceeds autopoetically, unpredictably but still in a rule-driven mode?

Artworks in the fields of dance, performance and video transfer swarm effects to their own individual mediality and materiality. This raises the question as to how

---

22 See also Carl Schmitt, *Die Theorie des Partisanen. Zwischenbemerkung zum Begriff des Politischen* (Berlin 1963).

swarms evince patterns of synchronizeted movements and manifest as participatory formations.

In this regard, Thomas Hauert's choreography *Accords* (2008) can be seen as a seminal example. The excerpt of the dance piece contains a group movement that slowly develops in relation to Maurice Ravel's *La Valse*: It starts with individual circles and lines of the dancers walking, running, slowing down, in constantly fluent movements in space. The movements seem contingent, but nevertheless there is a strong coherence within the group of the dancers. How is the process of synchronisation within a constantly changing and transforming flow of movement achieved? And how do the spatial figures of breaking away, getting closer and reaching out develop as a collective choreography? It is obvious that this kind of moving, of sharing the space and the timing of a rhythmic constellation, is *not* based on imitation. The emergence of a swarm-like dynamic and the coherence of this *corps* of dancers is part of the rhythmical processes of a spontaneous and unprescribed alignment. This *drive* of coherence is based on a specific training in kinaesthetic awareness: The dancers of Thomas Hauert's company ZOO are trained in a very intense investigation of bodily awareness. This includes the complexity of movement-*neighborhoods* and their temporal and spatial synchronisations and de-synchronisations, in a contingent exchange with the environment and the bodily displacements.

7.

The second example illustrates a different mode of *choreographing the swarm*: The performances of the Hamburg-based group LIGNA focus on movement-events in public spaces. They organize a *swarming of flash mobs* – collectives in public places like stations or squares (for example, train stations in Hamburg or in Leipzig).[23] But unlike Rheingold's examples, the collective movements of the LIGNA performances are not triggered by the use of smart phones, twitter or electronic media , but by an *older* and traditional medium: the *radio*. LIGNA relates to Walter Benjamin's writings on media, and therefore – somewhat ironically – call the type and structure of these public performances a "Radioballett". LIGNA composes a different type of *swarm*-effect and collective movement than Thomas Hauert's *Accords*. The radio works as a medium which induces the movements, spatial figurations and temporal synchronisation by clear and simple ver-

---

23 See also Kai van Eikels, "This Side of the Gathering. The Movement of Acting Collectivity: Ligna's *Radioballett*", *Performance Research, 13:1, On Choreography* (March 2008), Routledge, London: p. 85–98.

bal (audio-)directions – or, in other words, by external instructions. In a similar way, Janet Cardiff's performances are based on audio- and video-mappings or movements that guide the spectator or the audience through public spaces (like a theatre, or – at the *documenta* 2012 – through the Kassel railway station). Thus, the audience itself becomes a *performer* in the play. The paradox of this media transfer of collective movement constellations is the interstice between coherence (by synchronising the movement through media-directions) and dispersal. The placements and the corporeal neighborhoods are at the same time creating topographies of belonging (to)[24] and of exclusion: Who is participating in the movement of the swarm? Where are the borders of Inside and Outside? My research explores those effects of indecidability – with the specific consideration of the spatio-temporal intervals between the respective lines of action (Handlungen)[25] and participation. How do transitory collectives emerge within the framings of choreography? And where are the boundaries of such collectives which evoke feelings of inclusion or exclusion among individuals, or make them refuse to participate? The modes of coherence, of a feeling of belonging to, or – in contrast – to feel free to move within the dispersing crowd as a swarming collective necessitate a "rethinking the public sphere"[26] in terms of neighbors, of participation and of the space of sensing the Other.[27]

—

24 Compare Irit Rogoff, "The Implicated," paper presented at the conference *Performing the Future* at the House of World Cultures (Berlin, 9 July 2010).

25 See also Hannah Arendt, *The Human Condition* (Chicago 1998).

26 See also Nancy Fraser, "Rethinking the Public Sphere: A Contribution to the Critique of Actually Existing Democracy," ed. Craig Calhoun, *Habermas and the Public Sphere* (M.I.T. Press 1991), p. 109–142.

27 See also Homi K. Bhabha, *Our Neighbours, Ourselves: Contemporary Reflections on Survival* (Berlin 2011).

AUTHORS

Dr. ing. Henriette Bier After graduating in architecture (1998) from the University of Karlsruhe in Germany, Henriette Bier worked with Morphosis (1999–2001) in the USA on internationally relevant projects. She taught computer-based architectural design (2002–2003) at Universities in Austria, Germany and the Netherlands and worked on her Ph.D. at the Delft University of Technology (2004–08). Her Ph.D. research focused not only on analysis and critical assessment of digital technologies in architecture, but also reflected evaluation and classification of digitally-driven design through procedural- and object-oriented studies. It defined methodologies of digitally-driven design, which exploit Intelligent Computer-based Systems in order to support the design process as well as to actuate architecture. Since completing her Ph.D., she has been focusing on applications of robotics in architecture and has been teaching and researching as Assistant Professor at the Delft University of Technology in the Netherlands. She regularly lectures and publishes internationally.

Publications:

Henriette Bier, "Interactive Building," *Advances in Internet of Things*, Vol. 2, No. 4, AIT (2012).
Kas Oosterhuis and Henriette Bier, *Robotic(s in) Architecture* (Heijningen: Jap Sam Books, 2013)
Henriette Bier and Yeekee Ku, "Generative and Participatory Parametric Frameworks for Multi-player Design Games," *Footprint 13th issue*, ed. Maros Krivy and Tahl Kaminer (Delft: Stichting Footprint, 2013).
Henriette Bier and Terry Knight, "Dynamics of Data-driven Design," *Footprint 15th issue*, ed. Henriette Bier and Terry Knight (Delft: Stichting Footprint, 2014).

Prof. Dr. Gabriele Brandstetter is a Professor of Theater and Dance Studies at Freie Universität Berlin. Her research focus is on history and aesthetics of dance from the 18th century until today, theater and dance of the avant-garde; contemporary theater and dance, performance, theatricality and gender differences; concepts of body, movement and image. In 2004 she was awarded the Gottfried-Wilhelm-Leibniz-Preis of the German Research Foundation, through which she formed the Centre for Research on Movement at the Free University Berlin. Since 2007 Gabriele Brandstetter is co-director of the International Centre "Interweaving performance studies".

Publications:

Gabriele Brandstetter, *Tanz-Lektüren. Körperbilder und Raumfiguren der Avantgarde* (Freiburg: Rombach Verlag, 1995; extended by a third part ed. 2013).
Gabriele Brandstetter, co-editor *Methoden der Tanzwissenschaft. Modellanalysen zu Pina Bauschs 'Sacre du Printemps'* (Bielefeld: transcript, 2007).

Gabriele Brandstetter, *Schwarm(E)Motion. Bewegung zwischen Affekt und Masse* (Freiburg: Rombach Verlag, 2007).
Gabriele Brandstetter, co-editor *Touching and Being Touched. Kinesthesia and Empathy in Dance and Movement* (Berlin: de Gruyter, 2013).

Prof. Dr. Sándor Fekete studied mathematics and physics at the University of Cologne and earned his Ph.D. in Combinatorics and Optimization from the University of Waterloo, Canada (1992). After spending a year as postdoc at SUNY Stony Brook, he returned to Cologne, where he got his habilitation in mathematics (1998) and joined the optimization group at TU Berlin. In 2001 he became a professor of mathematics at TU Braunschweig; since 2007 he holds a newly-founded chair on algorithmics in the Computer Science Department in Braunschweig. He has published over 150 papers with more than 150 coauthors; his interests range all the way from theoretical foundations of algorithms and optimization to applications areas such as practical computer science, electrical engineering, economics, biology and physics.

Publications:
Sándor P. Fekete, B. Hendriks, C. Tessars, A. Wegener, H. Hellbrück, S. Fischer and S. Ebers. "Methods for Improving the Flow of Traffic," *Organic Computing - A Paradigm Shift for Complex Systems* (Basel: Birkhäuser Verlag, 2011), p. 447–460.
Sándor P. Fekete, C. Schmidt, A. Wegener, H. Hellbrück and S. Fischer, "Empowered by Wireless Communication: Distributed Methods for Self-Organizing Traffic Collectives," *ACM Transactions on Autonomous and Adaptive Systems* 5(3) (2010), p. 1–30.
Sándor P. Fekete, C. Schmidt, A. Wegener and S. Fischer, "Hovering Data Clouds for Recognizing Traffic jams," *Proceedings of 2nd International Symposium on Leveraging Applications of Formal Methods, Verification and Validation* (IEEE-ISOLA) (2006), p. 213–218.

Prof. Dr. Manfred Füllsack is Professor for Systems Sciences at the Karl Franzens University in Graz, concerned with modeling, computer-based simulation, network theory and evolutionary computation. Current research interests focus on the epistemology of computation, the sociology and economy of work and labor conditions, the diffusion of innovation, cooperation research and on the sociological conception of Niklas Luhmann. Studies of informatics, philosophy, mathematics, sociology and music at the University of Vienna. Several sojourns abroad including Michigan State University and the Sociological Institute of the Russian Academy of Sciences.

Publications:
Manfred Füllsack, *Networking Networks. Origins, Applications, Experiments. Proceedings of the Multi-disciplinary Network for the Simulation of Complex Systems – Research in the Von-Neumann-Galaxy* (Wien: Turia+Kant, 2013).

Manfred Füllsack, "Observing Productivity. What it Might Mean to be Productive, when Viewed through the Lens of Complexity Theory," *Journal for Philosophical Economics* VI/1 (2012), p. 2–23.
Manfred Füllsack, *Gleichzeitige Ungleichzeitigkeiten. Eine Einführung in die Komplexitätswissenschaften* (Wiesbaden: Springer, 2011).

Dr. Sebastian Gießmann is Academic Coordinator of the DFG Research Training Group "Locating Media" at the University of Siegen (Germany). His research interests include cultural techniques of cooperation, network history, material culture, anthropology of law, and Internet studies. He co-edits the *Zeitschrift für Kulturwissenschaften* and *ilinx*, the Berlin Journal in Cultural History and Theory. Sebastian Gießmann serves as co-speaker of the working group on data and networks in the German Society for Media Studies.

Publications:
Sebastian Gießmann, *Die Verbundenheit der Dinge. Eine Kulturgeschichte der Netze und Netzwerke* (Berlin: Kadmos, 2014).
Sebastian Gießmann, "Henry C. Beck, Material Culture and the London Tube Map of 1933," *Amodern 2, Network Archaeologies* (2013), ed. Nicole Starosielski, Braxton Soderman and Chris Cheek.
Sebastian Gießmann, "Verunreinigungsarbeit. Über den Netzwerkbegriff der Akteur-Netzwerk-Theorie," *Zeitschrift für Kulturwissenschaften* 1 (2013), p. 133–144, ed. Nacim Ghanbari and Marcus Hahn.

Prof. Dr. Tobias Harks earned his PhD in mathematics in 2007 at the Technical University Berlin where he also obtained his habilitation in 2012. Currently his research is funded by a Marie-Curie-Fellowship obtained in 2012. His main research interests include discrete optimization, algorithmic game theory and operations research. In 2011 he moved to Maastricht University as an Assistant Professor for operations research.

Publications:
T. Harks, M. Hoefer, M. Klimm and A. Skopalik, "Computing Pure and Strong Nash Equilibria in Bottleneck Congestion Games," *Mathematical Programming* (Ser. A), 141(1), p. 193–215 (2013).
P. von Falkenhausen and T. Harks, "Optimal Cost Sharing for Resource Selection Games," *Mathematics of Operations Research* 38(1), p. 184–204 (2013).
T. Harks and M. Klimm, "On the Existence of Pure Nash Equilibria in Weighted Congestion Games," *Mathematics of Operations Research* 37(3), p. 419–436 (2012).
T. Harks and K. Miller, "The Worst-Case Efficiency of Cost Sharing Methods in Resource Allocation Games," *Operations Research*, 56(6), p. 1491–1503 (2011).

Prof. Dr. Dirk Helbing is Professor of Sociology, in particular of Modeling and Simulation, and member of the Computer Science Department at ETH Zurich. He earned a Ph.D. in physics and was Managing Director of the Institute of Transport & Economics at Dresden University of Technology in Germany. He is internationally known for his work on pedestrian crowds, vehicle traffic, and agent-based models of social systems. Furthermore, he coordinated the FuturICT Initiative, which focused on the understanding of techno-socio-economic systems, using Big Data. Helbing is elected member of the World Economic Forum's Global Agenda Council on Complex Systems and of the German Academy of Sciences *Leopoldina*. He is also Chairman of the Physics of Socio-Economic Systems Division of the German Physical Society and co-founder of ETH Zurich's Risk Center. In 2013, he became a board member of the Global Brain Institute in Brussels. Within the ERC Advanced Investigator Grant "Momentum" he works on social simulations based on cognitive agents. His recent publication in *Nature* discusses globally networked risks and how to respond. He furthermore contributed to unveiling the hidden laws of global epidemic spreading. In 2014, he received a honorary Ph.D. from the TU Delft jointly from the Faculty of Technology, Policy and Management and the Faculty of Civil Engineering and Geosciences.

Publications:
Maximilian Schich, Chaoming Song, Yong-Yeol Ahn, Alexander Mirsky, Mauro Martino, Albert-László Barabási and Dirk Helbing, "A Network Framework of Cultural History," *Science* 345 (6196) (2014), p. 558–562.
Dirk Brockmann and Dirk Helbing, "The Hidden Geometry of Complex, Network-Driven Contagion Phenomena," *Science* 342 (6164) (2013), p. 1337–1342.
Dirk Helbing, "Globally Networked Risks and How to Respond," *Nature* 497 (2013), p. 51–59.

Prof. Dr. Martin Hoefer leads an Independent Reserach Group at Saarland University and is coordinator of the Algorithmic Game Theory Group at Max Planck Institute for Informatics, Saarbrücken. He received a Ph.D. in Computer Science from Konstanz University in 2007 and a Habilitation from RWTH Aachen University in 2011. His main research interests include design and analysis of algorithms, algorithmic game theory, and distributed computing.

Publications:
P. Berenbrink, M. Hoefer, T. Sauerwald, "Distributed Selfish Load Balancing on Networks," *ACM Transactions on Algorithms* 11(1), article 2, 2014.
T. Harks, M. Hoefer, M. Klimm, A. Skopalik, "Computing Pure and Strong Nash Equilibria in Bottleneck Congestion Games," *Mathematical Programming* (Ser. A) 141(1):193–215, 2013.
"Local Matching Dynamics in Social Networks," *Information & Computation* 222:20–35, 2013.
J. Dams, M. Hoefer, T. Kesselheim, "Convergence Time of Power-Control Dynamics," *IEEE Journal on Selected Areas in Communications* 30(11):2231–2237, 2012.

Dr. Shintaro Miyazaki studied Media Studies, Musicology and Philosophy at the University of Basel. In 2012, he finished his Ph.D. thesis on media archaeology of computation at Humboldt-University of Berlin. Around the same time he was a resident fellow at Akademie Schloss Solitude in Stuttgart. Since 2013, he researches and teches at the University of Applied Sciences and Arts Northwestern Switzerland, Academy of Art and Design, Institute of Experimental Design and Media Cultures in Basel, Switzerland. He is creating and designing epistemic devices, models and open-source software for inquiring media and their signals, timing and effects on history, culture and society.

Publications:

Shintaro Miyazaki, "Going Beyond the Visible: New Aesthetic as an Aesthetic of Blindness?," *The New Aesthetic* (forthcoming, 2015) ed. David M. Berry and Michael Dieter.
Shintaro Miyazaki, "AlgoRHYTHMS Everywhere – a Heuristic Approach to Everyday Technologies," *Pluralizing Rhythm: Music, Arts, Politics*, no. 26 (2013), p. 135–48, ed. Birgitte Stougaard and Jan Hein Hoogstad.
Shintaro Miyazaki, "Urban Sounds Unheard-of: A Media Archaeology of Ubiquitous Infospheres," *Continuum 27*, no. 4 (2013), p. 514–22.

Dr. Christina Vagt teaches cultural and media theories at the Institute of Philosophy, Literature, and the History of Science and Technology at the Berlin Institute of Technology. In 2013 she was visiting professor for media history and media theory at Bauhaus University Weimar, and in 2012 she was visiting fellow and Fulbright scholar at the Department of Comparative Literature at Stanford University. From 2005 to 2007, she was a DFG scholarship holder at the Graduate School Media of History – History of Media at Bauhaus-University Weimar, and from 2010 to 2011 Research Assistant (Post-Doc) at the Institute for Musicology and Media Studies, Humboldt-University Berlin. She is the author of *Geschickte Sprünge. Physik und Medium bei Martin Heidegger* (Berlin-Zürich: Diaphanes, 2012) and of numerous essays on cultural technologies, media, and aesthetics.

Publications:

Christina Vagt, "All Things Are Vectors. Kosmologie und Synergetik bei Richard Buckminster Fuller und Alfred North Whitehead," *Wissensgeschichte der Synergie*, ed. Tatjana Petzer and Stephan Steiner (Berlin: Zentrum für Literatur- und Kulturfoschung, forthcoming).
Christina Vagt, "Im Äther. Einstein, Bergson und die Uhren der Mikrobe," *Übertragungsräume. Medienarchäologische Perspektiven auf die Raumvorstellungen der Moderne* ed. Eva Johach and Diethard Sawicki (Wiesbaden: Reichert, 2013), p. 133–144.
Christina Vagt, editor of the first German Edition of Henri Bergson, *Dauer und Gleichzeitigkeit* (Hamburg: Philo Fine Arts, forthcoming).
Christina Vagt, co-editor *The Afterlives of Systems, Communication+1*, Vol.3, 2014.

Dr. Sebastian Vehlken is Junior Director of the Institute for Advanced Studies on Media Cultures of Computer Simulation (MECS), Leuphana University Lüneburg. He studied Media Studies and Economics at Ruhr-University Bochum and at Edith Cowan

University, Perth. From 2005–2007, he was a DFG scholarship holder in the Graduate School *Media of History – History of Media* at Bauhaus-Universität Weimar, and from 2007–2010 Research Associate (PreDoc) in Media Philosophy, University of Vienna. In 2010, he finished his Ph.D. thesis on a media history of biological and computational swarm research at the Institute for Culture of the Humboldt University Berlin. From 2010–2013, he was a Research Associate (PostDoc) at the Institute for Culture and Aesthetics of Digital Media, Leuphana University Lüneburg, and a Research Fellow at the Internationales Forschungskolleg Kulturwissenschaften, Vienna, in 2014. His current research project explores the media history of computer simulations in the context of civil nuclear energy technology in West Germany (FRG), with a focus on the development of fast breeder programs. Its preliminary title is *Plutonium Worlds. Computer Simulation and Nuclear Energy 1950–1980.*

Publications:

Sebastian Vehlken, "Plutonium Worlds. Fast Breeders, Systems Analysis and Computer Simulation in the Age of 'Hypotheticality'," *Communication+1, Vol. 3: Afterlife of Systems*, guest editors Florian Sprenger and Christina Vagt) (2014). URL: http://scholarworks.umass.edu/cpo/vol3/iss1/7
Sebastian Vehlken, "Zootechnologies. 'Swarming' as a Cultural Technique," *Theory, Culture and Society. Special issue Cultural Techniques*, ed. Geoffrey Winthrop-Young, Jussi Parikka and Ilinca Irascu (2012), p. 110–131.
Sebastian Vehlken, "Reality Mining. On New (and Older) Methods of Social Simulation," *New Masses – Social Media*, ed. Claus Pias, Timon Beyes and Inge Baxmann (Chicago: Chicago University Press (in press)).

Carolin Wiedemann, M.A., studied Journalism and Communication Studies and Sociology at the Sorbonne in Paris and at the University of Hamburg. Currently she is finishing her Ph.D. thesis that examines online collectivity phenomena and subversive networking strategies (e.g., *Anonymous*). Besides that she works as a freelance writer with publications including *Frankfurter Allgemeine Sonntagszeitung* and *Spiegel Online.*

Publications:

Carolin Wiedemann, "'Greetings from the Dark Side of the Internet' – Anonymous und die Frage nach Widerstand in Zeiten der Informatisierung," *Österreichische Zeitschrift für Soziologie*, Vol. 39. (2014).
Carolin Wiedemann, "Between Network, Swarm and Multitude. Anonymous and the Infrastructures of the Common," *Distinktion: Scandinavian Journal of Social Theory*, forthcoming, published online: http://www.tandfonline.com/doi/abs/10.1080/1600910X.2014.895768#.VCKnWOd3aco.
Carolin Wiedemann, "In den Ruinen der Repräsentation? Affekt, Agencement und das Okkurente," *FKW// Zeitschrift für Geschlechterforschung und visuelle Kultur, 55 - New Politics of Looking? Affekt und Repräsentation* (2014).